VEHICLE FUEL ECONOMY

VEHICLE FUEL ECONOMY

GOVERNMENT ACCOUNTABILITY OFFICE

Novinka Books
New York

Copyright © 2008 by Nova Science Publishers, Inc.

All rights reserved. No part of this book may be reproduced, stored in a retrieval system or transmitted in any form or by any means: electronic, electrostatic, magnetic, tape, mechanical photocopying, recording or otherwise without the written permission of the Publisher.

For permission to use material from this book please contact us:
Telephone 631-231-7269; Fax 631-231-8175
Web Site: http://www.novapublishers.com

NOTICE TO THE READER

The Publisher has taken reasonable care in the preparation of this book, but makes no expressed or implied warranty of any kind and assumes no responsibility for any errors or omissions. No liability is assumed for incidental or consequential damages in connection with or arising out of information contained in this book. The Publisher shall not be liable for any special, consequential, or exemplary damages resulting, in whole or in part, from the readers' use of, or reliance upon, this material.

This publication is designed to provide accurate and authoritative information with regard to the subject matter covered herein. It is sold with the clear understanding that the Publisher is not engaged in rendering legal or any other professional services. If legal or any other expert assistance is required, the services of a competent person should be sought. FROM A DECLARATION OF PARTICIPANTS JOINTLY ADOPTED BY A COMMITTEE OF THE AMERICAN BAR ASSOCIATION AND A COMMITTEE OF PUBLISHERS.

LIBRARY OF CONGRESS CATALOGING-IN-PUBLICATION DATA

Available upon request
ISBN 978-1-60692-057-2

CONTENTS

Correspondence		vii
Abbreviations		xi
Chapter 1	Results in Brief	1
Chapter 2	Background	5
Chapter 3	Several Weaknesses in the CAFE Program Exist	19
Chapter 4	Some Market-Based Policy Options Could Complement the CAFE Program	33
Chapter 5	Conclusions	47
Chapter 6	Matters for Congressional Consideration	49
Appendix I	Scope and Methodology	51
Appendix II	Selected Manufacturers' CAFE Performance, Selected Years from 1990 through 2005	55
References		57
Index		65

CORRESPONDENCE

August 2, 2007

The Honorable Daniel K. Inouye
Chairman
Committee on Commerce, Science, and Transportation
United States Senate

Dear Mr. Chairman:

Recent concerns over national security, environmental stresses, and economic pressures from increased fuel prices have led to a heightened interest in reducing oil consumption. For example, the President announced in early 2007, a nationwide goal to reduce gasoline consumption 20 percent from the levels that the administration projects would otherwise occur by 2017. Efforts to reduce oil consumption will need to include the transportation sector because transportation in the United States currently accounts for 68 percent of the nation's oil consumption, and cars and light trucks consume 60 percent of the oil consumed in the transportation sector.

In the aftermath of the energy crisis of the early 1970s—to reduce the nation's reliance on oil, a large part of which comes from other countries—Congress developed the Corporate Average Fuel Economy (CAFE) program for cars and light trucks. Under the CAFE program, manufacturers must ensure that the new vehicles in their fleets, on average, meet a specified miles per gallon (mpg) standard or pay a penalty.

The National Highway Traffic Safety Administration (NHTSA), an administration within the Department of Transportation (DOT), is primarily responsible for setting and enforcing CAFE standards, although the

Environmental Protection Agency (EPA) also plays a role in the program and the Department of Energy (DOE) is involved in setting national energy policy. Many changes in automotive technologies and the auto industry have occurred since the program was designed in the 1970s. These developments, along with the recent security, environmental, and economic concerns mentioned above have led to some changes in the CAFE program and to calls for further alterations, including raising CAFE standards or revising the way the program applies the standards. Several proposals to implement policies apart from the CAFE program would also attempt to increase vehicle fuel economy or reduce oil consumption through regulation, incentives, tax credits, or other means.

To assist Congress in addressing these issues, you asked us to discuss (1) how the CAFE program is designed to reduce oil consumption by cars and light trucks and the status of the program; (2) the strengths and weaknesses of the current CAFE program and NHTSA's capabilities to revise the program; and (3) other market-based policies—both existing and proposed—that are available to complement or possibly replace CAFE in reducing oil consumption by cars and light trucks and some strengths and weaknesses of these policies. To obtain information on how the CAFE program is designed, we reviewed U.S. Code and program guidance, including rule-making documents, and interviewed officials from federal agencies involved in the program, including NHTSA, EPA, and DOE. To obtain information about the strengths and weaknesses of the CAFE program and NHTSA's capabilities to further revise CAFE standards, we reviewed CAFE program budgets, key studies, and other documentation and interviewed NHTSA officials and experts in fuel economy and safety. We also interviewed the applicable automobile workers trade union (UAW), industry groups representing the automobile manufacturers, automotive safety experts, insurance industry representatives, and environmental advocates. To obtain information on other policy options for reducing oil consumption by cars and trucks, we reviewed published research and interviewed more than 30 experts in fuel economy from universities and advocacy organizations, and other industry stakeholders. We selected these experts in part by contacting officials who worked on a 2002 National Academy of Sciences (NAS) report on CAFE standards. During these conversations, we asked them to identify additional experts for us to contact. We also contacted officials in selected foreign countries with programs designed to reduce oil consumption for cars and light trucks. We did not, however, evaluate the costs and benefits of these alternatives, nor try to rank them in terms of overall effectiveness or efficiency. We conducted our work from August 2006 through June 2007 in accordance with generally accepted government auditing standards. See appendix I for further details on the scope and methodology.

ABBREVIATIONS

CAFE	Corporate Average Fuel Economy
CARB	California Air Resources Board
CBO	Congressional Budget Office
DOE	Department of Energy
DOT	Department of Transportation
EPA	Environmental Protection Agency
EPCA	Energy Policy and Conservation Act
GM	General Motors
GPS	Global Positioning System
GVWR	gross vehicle weight rating
INS	Immigration and Naturalization Service
IRS	Internal Revenue Service
MPG	miles per gallon
NHTSA	National Highway Traffic Safety Administration
NAS	National Academy of Sciences
SUV	sport utility vehicle
UAW	United Auto Workers
USPS	United States Postal Service
VMT	vehicle miles traveled

Chapter 1

RESULTS IN BRIEF[*]

The CAFE program is designed to reduce oil consumption by cars and light trucks by holding automobile manufacturers responsible for meeting or exceeding specified mpg standards for cars and light trucks and assessing penalties for manufacturers who do not meet those standards. NHTSA has overall responsibility for setting standards and administering the program. EPA collects information on vehicle models' fuel economy so that NHTSA can calculate the annual CAFE results for manufacturers' fleets. See table 1 for current CAFE standards.

Table 1. Current CAFE Standards

Model Year	Domestic Car CAFE Standard	Imported Car CAFE Standard	Light Truck CAFE Standard
2007	27.5 mpg	27.5 mpg	22.2 mpg

Source: NHTSA.

Manufacturers whose average mpg does not meet NHTSA's standards for each fleet in any given year will be subject to a penalty if they did not earn "credits" by exceeding the standards in the previous 3 years, or do not submit a plan to exceed the standards up to 3 years in the future. Also, manufacturers may increase their CAFE levels if they sell vehicles that can run on fuels other than gasoline. In terms of the status of the program, in 2003 NHTSA raised the light truck CAFE standard from 20.7 mpg in model year 2004 to 22.2 mpg in model year 2007. Subsequently, NHTSA restructured the CAFE program for light trucks using a method that

---

[*] This book is an excerpted indexed version of GAO Report GAO-07-921, Dated August 2007

categorizes light trucks based on their size and sets different standards for different sizes of light trucks beginning on a mandatory basis in model year 2011. However, NHTSA has not changed the CAFE standard for cars—27.5 mpg—since 1990, in part, because of provisions in DOT's appropriations acts for fiscal years 1996 through 2001 that prevented NHTSA from spending any funds to change CAFE standards. Recently, NHTSA officials stated that they wanted to restructure the car CAFE program before raising the car standard to avoid potential negative safety effects. However, NHTSA does not have the authority to restructure the program for cars. In 2007, as part of the administration's plan to meet the President's gasoline-reduction goal, the administration proposed legislation to Congress that would allow NHTSA to restructure the car CAFE program based on an attribute of the vehicle, such as size. This legislation is similar to NHTSA's recent changes to the light truck program. Several members of the 110th Congress have also introduced legislation to raise CAFE standards for cars and light trucks. In June 2007, the Senate passed a bill that would raise the CAFE standards for cars and light trucks and, among other things, allow a restructuring similar to that proposed by NHTSA. As of July 2007, the House has not acted on this bill.

According to estimates by the National Academy of Sciences (NAS) and other experts we consulted, the CAFE program has helped save billions of barrels of oil and could continue to do so in the future, but the program has several weaknesses and is not the only potential solution to reducing the nation's oil consumption over time. Several strengths make the CAFE program a viable and effective tool to help the nation meet its current oil-saving goals. First, as noted, many experts have concluded that CAFE has helped save oil—for example, a study by NAS[1] estimated that in 2002 CAFE contributed to saving 2.8 million barrels of fuel a day, or 14 percent of consumption in that year—and that increases to CAFE standards would contribute to future oil savings. NAS also stated that as of 2002, automakers could improve the fuel economy of most vehicle classes without large increases in vehicle costs. In addition, NHTSA's recent reform of the light truck program to a new attribute-based standard helped address safety, consumer choice, and manufacturer equity concerns. Through this reform, NHTSA was able to increase fuel economy standards for light trucks while also ensuring that CAFE was compatible with other important issues affecting cars and light trucks, such as safety. However, the CAFE program has several characteristics that hinder its effectiveness. For example, most manufacturers are already meeting or exceeding CAFE standards, so decisions by NHTSA and Congress not to raise the car CAFE standard since

1990 have reduced the incentive manufacturers have to increase the fuel economy of new cars. Furthermore, CAFE is not the most cost-effective[2] approach to reducing oil consumption. To further reduce the nation's oil consumption over time therefore may require more comprehensive and cost-effective approaches—some of which are discussed in the next section. Further, several refinements to the CAFE program could improve its effectiveness and make it less costly, such as instituting a CAFE credit-trading program to give manufacturers more flexibility in meeting the standards. The bill the Senate passed in June 2007 would institute an attribute-based CAFE system for cars and create a program where manufacturers could trade accrued CAFE credits with one another. Finally, our evaluation of NHTSA's capabilities suggests the agency could act quickly to implement new standards so CAFE standards could help the nation work toward reducing oil consumption in the immediate future.

Through reviews of our past reports and other studies, interviews with experts, reviews of recently proposed legislation, and analysis of existing programs in the United States and other countries, we identified several market-based policies involving cars and light trucks that could complement and strengthen the CAFE program's contribution to reducing oil consumption or that could serve as broader-reaching and potentially more cost-effective alternatives to CAFE. Market-based consumer incentives could complement CAFE by increasing consumer interest in purchasing vehicles with a high fuel economy. Several of these incentives already exist, such as the "Gas Guzzler Tax" on cars with a low fuel economy and tax credits for the purchase of fuel-saving hybrids. However, our review of these existing initiatives identified several limitations. Further we found other existing incentives that appear to work at cross purposes to those intended to reduce oil consumption, such as the relatively generous write-offs for purchases of sports utility vehicles for businesses. These incentives, if improved, could complement CAFE's fuel-saving effects; however, such incentives may not be enough to meet future goals for reducing oil consumption, even in conjunction with CAFE, because they are narrowly focused on influencing car purchases. Finally, market-based incentives to increase the availability and use of biofuels are being used to displace oil consumption.[3] However, our recent report on these efforts identified several limitations, and the cost-effectiveness of these programs is unclear.[4] Several options, such as a tax on fuel and carbon emissions or a carbon cap-and-trade program, provide incentives for consumers to engage in a number of fuel-saving behaviors. For example, increased gasoline taxes would likely influence consumers to reduce the amount of miles they drive in addition to purchasing fuel-efficient cars. In addition, such options could help the nation reach its oil consumption goals in a more cost-effective manner than the CAFE program. While these strategies could lead to larger reductions in oil consumption at lower cost to the nation, it would take time to design, garner support for, and implement each one.

This report includes matters for congressional consideration that, should Congress decide to increase fuel economy standards, it provide NHTSA with (1) express authority to reform the car CAFE program, (2) the resources to update information on new technologies, and (3) the flexibility to adjust the program in the future in response to changes in the passenger vehicle market. Also, to help ensure future CAFE standards are as affordable and effective as possible, we are recommending that NHTSA determine whether enhancements—including, but not limited to, credit trading, eliminating incentives to classify vehicles as light trucks, and indexing CAFE penalties to keep pace with inflation—should be made to the CAFE program. In addition, to ensure that existing and potential policies meant to reduce fuel consumption are achieving their goals, we are recommending that DOT, in cooperation with other relevant government agencies, evaluate what impact these policies are having or might have on fuel consumption. DOT, EPA, and DOE commented on a draft of this report. DOT officials generally concurred with the report's findings, did not believe indexing civil penalties to inflation would achieve further compliance with CAFE standards, and will consider the recommendations. Without more definitive research on the effect of increased penalties for not meeting CAFE standards, we continue to recommend that NHTSA consider studying the issue. EPA generally agreed with the report and recommendations and suggested we include more discussion on the issue of safety, which we did. Finally, DOE did not comment on the recommendations and did not agree with our finding that policy options other than CAFE, such as taxes and cap-and-trade programs, have the potential to produce fuel savings beyond what could be achieved through CAFE in a more cost-effective manner. We provided more information on the existing research we used to conclude that other approaches to reducing fuel use have the potential to be more cost-effective that the current program.

Chapter 2

BACKGROUND

Congress enacted the 1975 Energy Policy and Conservation Act (EPCA) during the aftermath of the energy crisis created by the Arab oil embargo of 1973 and 1974 to reduce oil consumption by the transportation sector in the United States.[5] EPCA established the CAFE program, which requires that manufacturers meet fuel economy standards for passenger cars and light trucks. To reduce oil consumption, the program uses fuel economy standards—measured in mpg—that cars and light trucks must meet separately. In addition to decreasing oil consumption by increasing the mileage driven on a gallon of gasoline, an increase in the standards also decreases some greenhouse gas tailpipe emissions.

EPCA established CAFE standards for passenger cars for model years 1978 through 1980 and 1985 and thereafter and gave NHTSA responsibility for administering the program and the authority to change the standards. However, the law prevents NHTSA from making structural reforms to the car CAFE program, such as basing the car CAFE standard on vehicle attributes such as size or weight. EPA also plays a role in the CAFE program. EPA implements testing procedures and tests vehicles to determine each model's fuel economy and determines the procedures for calculating the fuel economy values for CAFE for each manufacturer and for displaying the fuel economy levels on a new vehicle's window sticker.[6] The procedures for calculating fuel economy values are specified by the statute and include a separate test for city and highway fuel economy.

The standards called for manufacturers to produce passenger car fleets averaging 18 mpg in 1978, rising to 27.5 mpg by 1985.[7] In the 1980s, NHTSA reduced the CAFE standard for cars from 27.5 mpg to 26.0 mpg for model years 1986 through 1988, and to 26.5 mpg for model year 1989, in response to petitions from automakers who noted that consumers were demanding larger cars and engines, largely due to a decline in gasoline prices.

NHTSA issues new CAFE standards through a rule-making process. In the rule-making process, NHTSA issues a proposed rule and accepts comments from the public and stakeholders such as automakers, labor unions, and environmental advocacy groups. When determining what levels CAFE standards should be under an attribute-based system, as now exists for light trucks, NHTSA uses a cost-benefit model to determine the impact of various increases in CAFE standards on areas such as oil consumption and pollution. NHTSA must set standards at least 18 months before they take effect.

THE CAFE PROGRAM IS DESIGNED TO REDUCE OIL CONSUMPTION BY CARS AND LIGHT TRUCKS AND HAS BEEN RESTRUCTURED FOR LIGHT TRUCKS BUT NOT FOR CARS

To reduce oil consumption by light trucks and cars, NHTSA sets CAFE standards and levies penalties against manufacturers that do not meet the standards. In 2003, NHTSA raised light truck CAFE standards from 20.7 mpg in model year 2004 to 21.0 mpg in model year 2005, 21.6 mpg in model year 2006, and 22.2 mpg in model year 2007. Subsequently, NHTSA restructured the CAFE program for light trucks using a method that categorizes them based on their size and sets different targets for different sizes of light trucks to meet, beginning on an optional basis in model year 2008 and a mandatory basis in model year 2011. NHTSA has not raised the CAFE standard for cars above 27.5 mpg since 1990 because, among other reasons, NHTSA officials wish to first restructure the program to mitigate potential negative effects on safety of raising the standards. To that end, in 2007 the administration submitted a plan to restructure the program.[8] Several members of the 110th Congress introduced legislation to raise CAFE standards for cars and light trucks, and the Senate passed a bill in June 2007 increasing standards for cars and light trucks. The House had not acted on this bill as of July 2007.

NHTSA AND EPA IMPLEMENT A PRESCRIBED PROCESS TO ENSURE COMPLIANCE WITH FUEL ECONOMY STANDARDS

NHTSA determines a manufacturer's compliance with CAFE standards by comparing its fleet-wide fuel economy average against the appropriate CAFE standard.[9] Manufacturers, for their passenger car and light truck fleets, must meet separate CAFE standards, measured in mpg, or pay a penalty. In addition, manufacturers must separately meet CAFE standards for their imported and domestic passenger car fleets.[10] NHTSA defines light trucks as vehicles that are designed to perform functions such as carrying cargo, having an open-bed, carrying more than 10 passengers, or operating off-road.[11] Sport utility vehicles (SUV), short-bed pickup trucks, and passenger vans with a gross vehicle weight rating (GVWR) between 8,500 and 10,000 pounds have been considered medium-duty vehicles, and NHTSA has excluded them from the CAFE program until model year 2011, when NHTSA will include them in the CAFE program as light trucks.[12] Vehicles with a GVWR over 10,000 pounds are considered heavy-duty vehicles and are not subject to the CAFE requirements.

EPA allows manufacturers to test their own vehicles to determine their fuel economy, but EPA tests a sample of new vehicles at its National Vehicle and Fuel Emissions Laboratory in Ann Arbor, Michigan, to confirm the manufacturers' results. EPA reports the yearly CAFE results for each manufacturer to NHTSA for CAFE enforcement. NHTSA then determines if the manufacturers comply with the CAFE standards and assesses civil penalties against manufacturers who do not meet the standards. Compliance with the standards is measured by calculating a sales-weighted harmonic mean of the fuel economies of a given manufacturer's product line, with domestically produced cars, imported cars, and all light trucks measured separately. A manufacturer whose CAFE level for its passenger car or light truck fleet does not meet the standard for a given model year is subject to a civil penalty of $5.50 per tenth of a mpg that the manufacturer's CAFE level is below the required CAFE level multiplied by the number of vehicles in the affected fleet manufactured for a given model year. NHTSA collected more than $678 million in civil penalties from model years 1983 through 2005—mostly from European manufacturers producing high-performance, luxury vehicles.[13] Asian and domestic manufacturers have historically not paid penalties because they have either met or exceeded passenger car and light truck fleet CAFE requirements. See table 1 for a list of CAFE penalties paid, by manufacturer, for 2001 through 2005.

Table 2. CAFE Penalties by Manufacturer
Model Years 2001 through 2005

Model year	Manufacturer	Passenger car import penalty	Light truck penalty	Passenger car import penalty in 2006 dollars	Light truck penalty in 2006 dollars
2001	Volkswagen	$0	$173,118	$0	$196,159
2001	Porsche	4,997,190	0	5,662,281	0
2001	BMW	27,985,925	1,497,991	31,710,655	1,697,363
2001	Fiat	817,443	0	926,239	0
2001	Lotus	35,744	0	40,501	0
2002	Porsche	4,357,782	0	4,845,053	0
2002	BMW	14,066,124	0	15,638,947	0
2002	Fiat	1,344,222	0	1,494,528	0
2002	Lotus	36,850	0	40,970	0
2003	Ferrari/Maserati	1,139,710	0	1,242,024	0
2003	Porsche	3,348,609	189,635	3,649,221	206,659
2003	BMW	8,861,776	1,676,752	9,657,318	1,827,278
2004	Ferrari/Maserati	1,511,125	0	1,605,263	0
2004	Porsche	3,225,453	3,171,564	3,426,387	3,369,141
2004	Volkswagen	0	3,474,372	0	3,690,813
2004	DaimlerChrysler	8,537,364	0	9,069,212	0
2005	BMW	2,975,496	0	3,067,446	0
2005	DaimlerChrysler	16,895,472	0	17,417,585	0
2005	Ferrari/Maserati	2,426,413	0	2,501,395	0
2005	Porsche	2,238,082	0	2,307,244	0
2005	Porsche	0	1,977,250	0	2,038,352
2005	Spyker	3,157	0	3,255	0
2005	Volkswagen	0	1,136,668	0	1,171,794

Source: NHTSA.
Note: No manufacturers of domestic passenger cars needed to pay penalties during this period.

Another penalty that manufacturers might pay for producing car models that have low fuel economy levels is the so-called "Gas Guzzler Tax."[14] EPA reports the fuel economy test results for each manufacturer to the Internal Revenue Service (IRS), which imposes a tax on manufacturers of new model year cars that fail to meet a fuel economy level of 22.5 mpg. IRS collects the tax from the manufacturer after production has ended for the model year. The amount of the tax paid is displayed on a new vehicle's fuel economy window sticker. Although related, the Gas Guzzler Tax is not part of the CAFE program. Gas Guzzler Tax revenues are deposited into the Treasury, like CAFE penalties. Light trucks are not subject to the Gas Guzzler Tax.

Apart from paying penalties, manufacturers have another option if they do not comply with the CAFE standards in one model year—using so-called CAFE credits earned in other model years. For example, when the average fuel economy of either a manufacturer's passenger car or light truck fleet for a particular model year "overcomplies," or exceeds the established standard, the manufacturer earns credits it can use to make up a deficit in another model year.[15] These surplus credits can be applied to a deficit in any of the 3 consecutive model years immediately prior to or subsequent to the model year in which the credits are earned. Manufacturers must use any credits within 3 years of earning them. If a manufacturers has a deficit, but no (or not enough) credits available, the manufacturer can either pay the penalty or submit a plan to NHTSA on how the manufacturer will make up the deficit by earning a sufficient amount of credits in the next 3 years. NHTSA officials stated there is no express authority for trading credits between manufacturers, or for a manufacturer to transfer credits among different classes of a manufacturer's fleets (such as between cars and light trucks).

In addition, the Alternative Motor Fuels Act of 1988 gave credits to manufacturers for producing vehicles that could run on alternative fuels in addition to gasoline.[16] Under this so-called "Dual Fuel" program, manufacturers may increase their CAFE by up to 1.2 mpg for vehicles through model year 2010 that are capable of using both regular gasoline and an alternative fuel.[17]

NHTSA RECENTLY INCREASED STANDARDS AND REFORMED THE LIGHT TRUCK CAFE PROGRAM TO HELP ADDRESS DECLINING FUEL ECONOMY

NHTSA recently increased standards for and reformed the light truck CAFE program. The impact of the light truck market on overall oil consumption in the United States has grown since the beginning of the CAFE program as market share for these vehicles has increased. For example, in 1980, shortly after the program began, light trucks composed about 20 percent of the new passenger vehicle market in the United States. By 2005, light trucks, including minivans, pickup trucks, and sport utility vehicles, accounted for about 50 percent of the new passenger vehicle market in the United States. The overall fuel economy of the U.S. vehicle fleet declined in the 1990s, in part due to the increased market share of light trucks. (See fig. 1 showing share of fleet composed by light trucks).

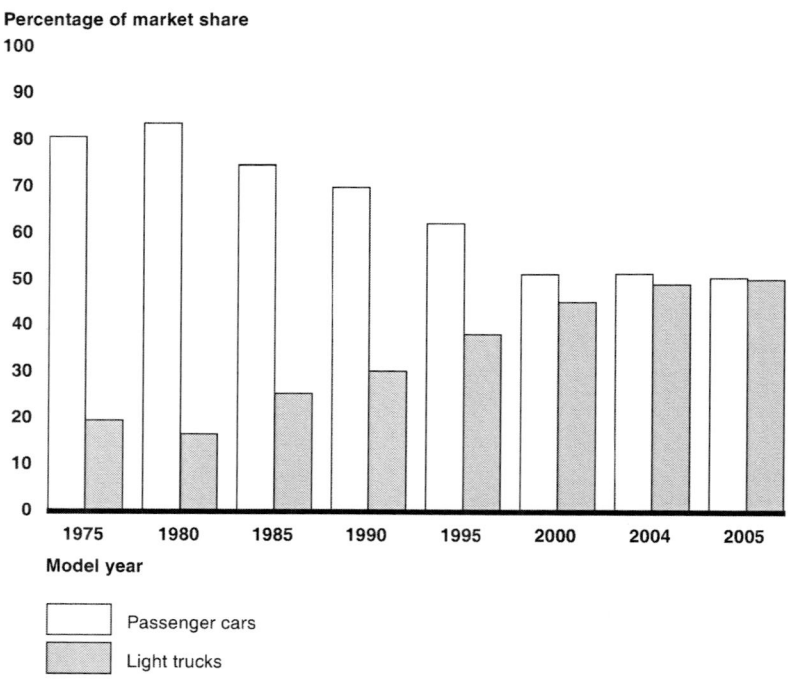

Source: GAO analysis of data from DOE/Transportation Data Energy Book, edition 25.

Figure 1. Increased Share of Light Trucks in the U.S. Passenger Vehicle Market.

To help address the overall declining fuel economy of the U.S. passenger vehicle fleet, in April 2003, NHTSA promulgated a final rule increasing light truck CAFE standards from 20.7 mpg in model year 2004 to 21.0 mpg in model year 2005, 21.6 mpg in model year 2006, and 22.2 mpg in model year 2007. In addition, the agency began investigating the possibility of reforming the light truck CAFE program in part to address safety concerns. The 2002 NAS report on the impact of CAFE standards stated that because the lowest cost way for an automobile manufacturer to increase vehicle fuel economy is to decrease vehicle weight, increases to CAFE standards—under the original CAFE system currently still in use for cars—could adversely affect safety and result in more highway fatalities.[18] The report also stated that past increases in CAFE standards had likely contributed to additional highway deaths, though other factors were also involved. The report recommended that NHTSA investigate implementing a new CAFE system based on the attributes of a vehicle.[19]

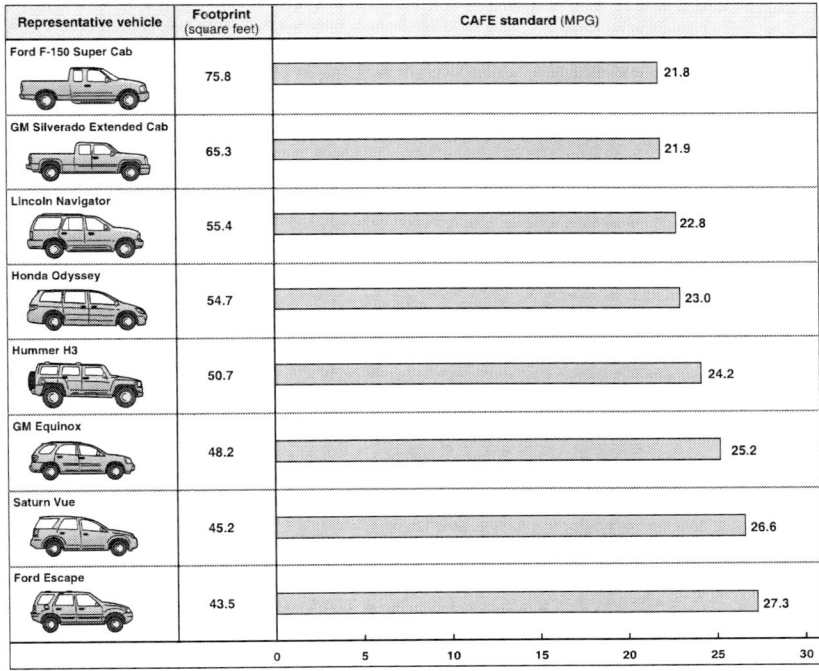

Source: GAO.

Figure 2. Application of Footprint-Based Light Truck CAFE Standards to Light Trucks of Different Sizes for Model Year 2011.

NHTSA issued a final rule in April 2006 that restructures the CAFE program for light trucks and continues to increase light truck CAFE standards for model years 2008 through 2011. Under the new rule, fuel economy standards are established based on truck size instead of having one average standard for all light trucks produced by a manufacturer. Each truck is assigned a fuel economy target based on a measure of vehicle size called "footprint," the product of multiplying a vehicle's wheelbase (the distance from front to the rear axles) by its track width (the horizontal distance between the tires). (See fig. 2 for a display of how the standard applies to trucks of different sizes). By model year 2011, all manufacturers will be required to comply with the reformed, footprint-based CAFE standard with a range of 21.8 mpg for the largest footprint trucks to 30.4 mpg for the smallest footprint trucks. NHTSA estimates that under the footprint-based system, light trucks will average 24.0 mpg in model year 2011. To facilitate the transition to the new system, NHTSA set both reformed and unreformed standards for model years 2008 through 2010 and manufacturers may choose to meet either standard during those years.

According to NHTSA officials, the footprint-based CAFE approach may enable the country to achieve larger reductions in oil consumption while enhancing safety.[20] Under the old standard, manufacturers who build a relatively larger share of smaller light trucks may already exceed the fleet CAFE standard and, therefore, would have little incentive to continue increasing the fuel economy of their light trucks. However, under the footprint-based standards, the required overall fuel economy of the light truck fleet will rise over time, since NHTSA has stated the targets will rise over time. NHTSA officials told us they believe this approach will spread the regulatory cost burden for fuel economy improvements more broadly across the industry instead of concentrating it more exclusively on the manufacturers of heavier, lower fuel economy vehicles. In addition, the footprint-based standards include some larger vehicles such as sport utility vehicles, with a GVWR between 8,500 and 10,000 pounds that previously were excluded from the CAFE program. NHTSA estimates that including these vehicles in the CAFE program will save 251 million gallons of fuel over the life of the vehicles sold in 2011.[21] In addition to these expected fuel savings, the footprint-based CAFE standards offer enhanced safety by discouraging downsizing of vehicles since, as vehicles become smaller, the applicable fuel economy target becomes more stringent.

NHTSA HAS NOT CHANGED THE CAR CAFE STANDARD SINCE 1990 BUT HAS REQUESTED AUTHORITY TO REFORM THE PROGRAM

NHTSA has not changed the car CAFE standard since 1990, but it has requested authority to reform the program so that it can raise the standard in the future. After reducing the 27.5 mpg car CAFE standard for model years 1986 through 1989, NHTSA raised it back to 27.5 mpg for the 1990 model year, and the standard has remained at 27.5 mpg since then. NHTSA officials cited several reasons for not raising the car CAFE standard over 27.5 mpg. First, for 6 years, Congress specifically prevented NHTSA from adjusting the CAFE standards. Beginning in fiscal year 1996 and lasting through fiscal year 2001, Congress included language in DOT's appropriations acts preventing NHTSA from expending any appropriated funds for rule makings to adjust CAFE standards for either cars or light trucks. Second, although NHTSA officials state that the agency has the legislative authority to raise the CAFE standard for cars above the 27.5 mpg standard, specified by the law, these officials stated that the law does not provide NHTSA with express authority for restructuring the program by, for example, developing a size-based standard for cars as it recently did for light trucks.[22] NHTSA officials stated they are reluctant to raise the car standard without also restructuring the program because they are concerned that increases in the car CAFE standard under the existing program could have a negative impact on safety by giving auto manufacturers an incentive to reduce the weight of the vehicles they build in order to meet increased fuel economy standards. NHTSA officials pointed out that, according to the 2002 NAS report, reducing the weight in vehicles may make vehicles less crashworthy and lead to increased highway fatalities.[23]

In 2007, the administration submitted proposed legislation to Congress that, if enacted, would give the Secretary authority to restructure and increase the CAFE standard for cars. The proposal calls for the continuation of the current statutory requirement that fuel economy standards be set at the maximum level that NHTSA believes the manufacturers could achieve in a specific model year. The proposal would also give NHTSA the authority to base the standard on one or more vehicle attributes, such as size, similar to the light truck standard, so that there would be different targets for cars with different attributes. Since product mix typically differs from manufacturer to manufacturer, each manufacturer would likely be subject to a unique CAFE requirement for its car fleet. In addition, the proposal calls for a credit trading system among manufacturers. If a manufacturer exceeds the mileage standard, it could sell its credits to another manufacturer or a third-party broker. The proposal does not provide a specific goal or mpg standard, but, as for the light truck standard, calls for setting a fuel economy standard that is the maximum feasible average fuel economy level that NHTSA decides the manufacturers can achieve in a specific model year.

In addition to this proposed legislation, several Members of Congress submitted bills that have some similarities to the Secretary's proposal but, if enacted, would set a minimum fuel economy standard for manufacturers to meet. For example, the Senate passed a bill that calls for cars and light trucks to achieve a combined CAFE average of 35 mpg by 2020.[24]

The CAFE Program Has Saved Billions of Barrels of Oil, but Car Standards Have Not Changed for Decades

According to estimates by NAS, the CAFE program has contributed to saving billions of barrels of oil and could continue to do so in the future, but several weaknesses in the program exist. Experts and industry stakeholders with whom we spoke generally attributed this success to the fact that CAFE was a mandatory standard, unlike voluntary standards in many other nations. Also, most of these experts and stakeholders agreed that NHTSA's recent reforms to the light truck CAFE program enhanced the program by reducing incentives for manufacturers to make vehicles less safe to meet CAFE standards and making the program more equitable for all manufacturers. In addition, experts and stakeholders cited the program's unintended effect of reducing greenhouse gas emissions as a strength of the program. However, the program has not kept pace with consumer preferences for larger vehicles, technology, or growing concern about fuel economy. These experts and stakeholders also cited several weaknesses in the program, noting that there are other cost-effective strategies to reduce oil consumption, the program is not automatically reviewed and adjusted over time, the program has not had its penalty structure changed since 1997, and the program—under the dual fuel program—gives CAFE credits to manufacturers who build vehicles that can run on alternative fuels, regardless of whether the drivers actually use those fuels. Our evaluation of NHTSA's capabilities and the agency's recent reform of the light truck program suggest that the agency generally has the capabilities to reform standards and could act quickly in the future to reform the car program if the necessary authority is provided. However, some of NHTSA's capabilities could be improved, such as increasing staff levels and updating data on fuel-efficient technology for use in its cost-benefit analysis.

THE CAFE PROGRAM HAS SEVERAL STRENGTHS, INCLUDING SAVING OIL; COMPATIBILITY WITH OTHER ISSUES, SUCH AS SAFETY OF CARS AND LIGHT TRUCKS; AND SLOWING THE INCREASE IN GREENHOUSE GAS EMISSIONS

Experts, NHTSA officials, and representatives from auto manufacturers with whom we spoke cited several strengths of the CAFE program. Most of these experts said CAFE was somewhat effective in reducing fuel consumption, and a study by NAS estimated that in 2002 CAFE, along with other factors, contributed to saving about 2.8 million barrels of fuel per day, or about 14 percent of consumption in that year. Many experts thought that CAFE's effectiveness is largely derived from introducing a mandatory standard that all auto manufacturers had to meet, unless the manufacturer was willing to pay a penalty. Compared with programs in other nations that have voluntary fuel economy standards the CAFE program's enforceable, mandatory standards have achieved favorable results though, in many of those countries, high fuel taxes and high fuel prices, especially in Europe, have reduced the need for fuel economy standards. In addition, according to NHTSA officials citing NAS results, the program has had a demonstrable record of increasing fuel economy in passenger cars and light trucks. These officials said they had concluded that if the CAFE program did not exist, auto manufacturers would produce less fuel-efficient cars than they currently produce. For example, before Congress established CAFE and set the standard for cars, there was no minimum standard for fuel economy in the United States. Between model years 1967 and 1974, the average domestic passenger car's fuel economy dropped from 14.8 mpg to 12.9 mpg. From model year 1978, when CAFE was first imposed, domestic passenger car fuel economy increased from 18.7 mpg to 30.0 mpg in 2004.[25] In the future, NHTSA officials stated they could further enhance fuel savings beyond what could be expected from the current CAFE program, with its single standard for all cars, by requiring fuel economy increases across a wide range of vehicles with different attributes, such as size, if they receive the authority to do so. As noted, NHTSA has made this change to the light truck program; and as of model year 2011, light trucks will be required to meet size-based CAFE standards, and the agency would like to institute a similar change for cars. The 2002 NAS report stated that the technology exists to increase fuel economy without large increases in vehicle costs.[26]

Manufacturers with whom we spoke agreed, though they preferred incremental increases to CAFE standards to ensure they could adjust to any new standards over aggressive CAFE increases over a short-term period.

According to NHTSA and several experts with whom we spoke, NHTSA's actions to reform the light truck standard allowed the agency to increase fuel economy standards for light trucks while also ensuring that CAFE was compatible with other important issues affecting cars and light trucks, including the following:

- *Enhancing Safety*: According to NHTSA officials and several experts with whom we spoke, the new size-based standard for light trucks removes the incentive for manufacturers to comply with CAFE by pursuing strategies that entail safety risks associated with increased highway deaths, such as downsizing vehicles and designing some vehicles to be classified as light trucks rather than cars, which may increase the vehicle's propensity to roll over. According to NHTSA, the size-based approach enables NHTSA to increase standards without encouraging these safety risks. For example, the approach does not provide incentives for manufacturers to downsize vehicles because smaller vehicles must meet more stringent CAFE standards.
- *Reflecting Consumer Choice*: NHTSA officials also stated that the attribute-based light truck CAFE program addresses some concerns about consumer choice. For instance, under the previous system, instead of installing more fuel-saving technologies across their fleets, manufacturers might have moved toward building fewer large vehicles and more smaller vehicles to meet new CAFE standards, even though consumers typically have not demanded them. In the attribute-based system, manufacturers must improve the fuel economy of all light trucks, no matter their size. As a result, according to NHTSA, manufacturers can continue to build a greater range of vehicles of varying sizes.
- *Creating a More Equitable Regulatory Framework*: The attribute-based standard also addresses concerns that raising CAFE standards in the previous system would tend to require only those manufacturers that produce a relatively larger share of light trucks to increase fuel economy in their vehicles to comply with a new standard, which places most of the cost and

compliance burdens on manufacturers that make a wide range of vehicles, including larger vehicles. Under the attribute-based system, however, NHTSA officials stated that it is more likely that additional manufacturers would have to increase the fuel economy of at least some of their vehicles in order to meet the new, size-based light truck CAFE standard. Most experts with whom we spoke agreed that additional manufacturers would have to increase fuel economy under the reformed system.

In addition to these strengths, the CAFE program has had the additional, positive, impact of slowing the rate of increase in transportation-related greenhouse gas emissions. A link exists between the amount of fuel burned and the growing amount of greenhouse gases in the atmosphere, which many agree contributes to global climate change. When the CAFE program increased fuel economy standards, it reduced greenhouse gas emissions from passenger cars and light trucks because as fuel economy is increased, the reduction in gasoline consumption translates into a reduction in carbon dioxide emissions. The transportation sector accounted for 27 percent of U.S. greenhouse gas emissions in 2003. EPA estimates that cars and light trucks account for 62 percent of the transportation sector's greenhouse gas emissions, as shown in fig. 3. Congress has pending several bills that would increase CAFE standards, in part, to reduce greenhouse gas emissions that are linked to climate change. Additionally, the California Air Resources Board (CARB) plans to implement fuel economy regulations for cars and light trucks sold in California that would exceed current CAFE standards, and several other states have announced similar plans, if EPA grants them the authority to do so.[27]

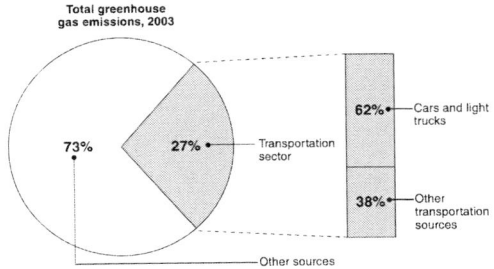

Source: EPA.

Figure 3. Sources of Greenhouse Gas Emissions in the Transportation Sector, 2003.

Chapter 3

SEVERAL WEAKNESSES IN THE CAFE PROGRAM EXIST

Despite the strengths of the CAFE program, experts and industry stakeholders with whom we spoke said aspects of the program were outdated and the program has not been revised to recognize or accommodate changes in technologies, consumer demand or the economics of the auto industry that have occurred since the program took effect in 1978. A longer discussion of these cited weaknesses follows:

- *Fuel economy standards have been allowed to stagnate*: The car CAFE standard has remained stagnant for nearly 2 decades. Meanwhile, there have been increases in the market share of larger vehicles, with relatively lower fuel economy ratings such as SUVs and minivans. Since the car CAFE standard returned to 27.5 mpg for the 1990 model year, the number of vehicle miles traveled on U.S. roads has also increased by about 31 percent, and the market share of light trucks has increased from about 20 percent to about 50 percent of the new vehicle fleet, resulting in more miles traveled by light trucks. This increase in the use of light trucks, along with consumers' preferences for higher performance vehicles, which generally achieve lower fuel economy than lower performance vehicles, has resulted in the overall fuel economy of the fleet declining from a high of 26.2 mpg in 1987 to 24.6 mpg in 2004, though the fleet fuel economy increased to 25.2 in 2005. Historically low gasoline prices over much of the last 2 decades have compounded this weakness, according to an expert with whom we spoke, since these low prices gave consumers little incentive to demand vehicles with higher fuel economy. However, two recent studies stated that the recent increase in gasoline prices is showing that consumers may be willing to pay more for fuel-efficient vehicles than in the past. One of the studies also cited consumers' growing concern about climate change as another reason to consider vehicles with a higher fuel economy. However, the level of emphasis consumers will be willing to place on these concerns remains to be seen and depends, in part, on the future level of gasoline prices.[28] During the time the car CAFE standard has remained stagnant, the industry average has met or exceeded the standard consistently. (See app. II for a description of selected manufacturers' CAFE performance since 1990.)

- *Lower-cost policies could achieve the goals of the program*: Although the CAFE program has contributed to reduced fuel consumption by cars and light trucks in the past and would continue to do so in the future, recent research and the experts with whom we spoke indicate that CAFE standards are not the most cost-effective option available. For example, studies done by GAO, the Congressional Budget Office (CBO), and others have found that the same fuel- saving goals could have been reached at lower cost to society if a more flexible policy that directly increased the cost of using these fuels or other petroleum products had been adopted. CBO has noted that the CAFE program could be made less costly and more effective than it currently is by instituting, for example, a broader credit trading program. However, other options, several of which are discussed later in this report, would also offer a less costly and more effective approach than the CAFE program.

- *The program's distinction between foreign and domestic cars is complicated and costly and may no longer be relevant*: Some experts also cited the distinction the CAFE program draws between foreign and domestic cars as a weakness in the program. Since the creation of the CAFE program, many domestically based manufacturers have begun to produce vehicles abroad, and many foreign manufacturers have begun to produce vehicles in the United States. For example, more than half of all the vehicles sold by foreign manufacturers in the

United States are produced in the United States. Also, because of the North American Free Trade Agreement, NHTSA treats vehicles made in Mexico or Canada as part of a manufacturer's domestic fleet. Several experts cited the distinction that the CAFE program is required to make between foreign and domestic cars as an outdated facet of the program that simply makes it more complicated or costly for auto manufacturers to comply with CAFE standards by adding more factors for manufacturers to consider when deciding about where to produce their vehicles. NHTSA officials stated they abolished the foreign and domestic vehicle distinctions they had created in the light truck CAFE program beginning in the 1996 model year in part because manufacturers were importing almost no light trucks into the United States. However, NHTSA has no authority to remove this distinction from the car CAFE program, and the administration did not request this authority in its proposal to Congress to grant NHTSA authority to reform the car CAFE program. Auto manufacturers and experts with whom we spoke supported abolishing this distinction but the UAW—the labor union that represents most workers at U.S.-owned auto manufacturers—opposes this, stating that the distinction gives manufacturers an incentive to produce all types of vehicles, including small vehicles, in the United States and expressing concerns that abolishing the distinction would result in auto manufacturers moving U.S. auto manufacturing jobs overseas. However, the 2002 NAS report reviewed this issue and found no perceptible effect on auto industry employment because of this distinction in the CAFE program.

- *Penalties may not be a strong deterrent as they have not increased since 1997*: Several experts with whom we spoke noted that penalties for violating CAFE standards have not increased since 1997, when, pursuant to the Federal Civil Penalties Inflation Adjustment Act of 1990, NHTSA raised the penalty from $5 to $5.50 per vehicle for every 0.1 mpg (or $55 per 1 mpg) by which a manufacturer's fleet falls short of the CAFE standard.[29] Several experts stated that this is not enough of a monetary incentive for manufacturers to comply with CAFE. For example, 22 manufacturers paid penalties during model years 1983 through 2005 (see table 3), including 5 companies that paid penalties 10 times. However, several experts also recognized that many auto manufacturers attempt to comply with CAFE standards more to avoid the negative public relations impact of not complying with CAFE standards than the actual financial penalty. Representatives of the domestic auto manufacturers confirmed this interpretation. A number of foreign manufacturers with whom we spoke stated that the civil penalty provisions in the law for failing to meet CAFE standards present a deterrent because they mean a violation is "unlawful conduct." These manufacturers believe it is an unacceptable business practice to plan to routinely fail to meet standards. Also, NHTSA staff told us the agency has not analyzed how the penalty structure could be modified to achieve higher compliance rates among foreign manufacturers that currently do not meet CAFE standards, but they noted that generally the manufacturers that pay penalties are manufacturers of luxury or specialty high-performance vehicles. NHTSA staff believes that as the sales of those vehicles are significantly dependent on their current level of performance, raising the penalty would not be likely to induce these companies to produce more fuel-efficient vehicles. Rather, NHTSA staff said that, in their opinion, customers of these vehicles would absorb the cost of higher penalties.

Table 3. Total CAFE Penalties Paid by Individual Manufacturers, 1983 through 2005

Manufacturer	Penalty in nominal dollars
Aston Martin Lagonda, Ltd.	$2,550
Autokraft, Ltd.	2,590
BMW of North America, Inc.	225,531,779
Consulier Industries	150
DaimlerChrysler Corp.	25,432,836
Ferrari Maserati North America, Inc.	5,077,248
Fiat Motors of North America, Inc.	10,791,076
Jaguar Cars, Inc.	40,069,650
Lotus Cars USA, Inc.	239,934
Maserati Automobiles of America, Inc.	121,600
Mercedes-Benz USA, LLC.	226,128,170
Panoz Auto Development Corp.	26,918

Manufacturer	Penalty in nominal dollars
PAS, Inc.	294,500
Peugeot Motors of America, Inc.	2,855,205
Porsche Cars North America, Inc.	52,437,258
Rover Group, Ltd.	23,092,226
Sterling Motor Cars	4,309,780
Spyker	3,157
Sun International	45
Vector Aeromotive Corp.	4,350
Volkswagen of America, Inc.	5,461,528
Volvo Cars of North America	56,421,280
Total	**$678,303,827**

Source: NHTSA.
Note: Amounts rounded to the nearest dollar and not adjusted for inflation.

- *NHTSA does not have authority to revise the car CAFE program according to vehicle attributes*: NHTSA's recent revision of the light truck CAFE program generally addressed safety and equity concerns; however, these concerns have not been addressed in the car CAFE program. NHTSA officials stated that one reason the agency had not increased the car CAFE standard is that under the current system, manufacturers may have an incentive to meet higher CAFE standards, primarily by making vehicles lighter and thus increasing their fuel economy. NHTSA told us it would prefer to institute an attribute-based standard for cars but does not have the authority. Officials also said that they were mindful of the 2002 NAS report, which stated that increases in CAFE standards in the late 1970s and early 1980s contributed to additional highway fatalities when manufacturers built smaller, lighter vehicles to meet the higher CAFE standards. One reason NHTSA reformed the light truck standard is to avoid having such adverse safety consequences again when raising CAFE standards for light trucks. However, NHTSA does not have the legal authority to revise the car CAFE program to implement a system of attributes that would include increases over time. Instead, it must use a single number for the entire fleet, though the administration has several times requested that Congress

provide such authority, and Congress is now considering these requests.
- *The current CAFE program for cars may create competitive advantages for certain manufacturers*: According to some experts with whom we spoke, the current car CAFE standard creates a competitive advantage for some auto manufacturers. For example, manufacturers that responded to growing consumer demand for larger vehicles by selling large sedans or SUVs must work harder and devote more of their resources to comply with CAFE because the larger vehicles lower their fleet fuel economy average. However, these vehicles are often among manufacturers' most profitable to sell. Manufacturers whose sales are focused mostly on smaller vehicles, which tend to have relatively higher fuel economy due to their relatively low weight, have less incentive to further use their expertise and install more fuel-saving technologies and do not have to spend resources attempting to meet higher standards. As a result, the manufacturers may be able to spend those resources on developing new models, marketing, or other activities giving them an advantage. However, raising the CAFE standard by instituting an attribute system requires all manufacturers to increase the fuel efficiency of their vehicles.
- *The Dual Fuel program allows manufacturers to achieve a lower fuel economy than otherwise would be required under CAFE*: The Dual Fuel program, which was established by the Alternative Motor Fuels Act of 1988, provides for auto manufacturers an opportunity to increase their CAFE rating in return for producing flex-fuel vehicles capable of running on conventional gasoline or alternative fuels (typically an ethanol blend known as E85). The program was designed to encourage development and increased the availability of alternative fuels by creating a market for these fuels by giving manufacturers an incentive to build vehicles that could run on them. EPA and NHTSA officials with whom we spoke estimated that adding equipment to make vehicles capable of using alternative fuels in addition to gasoline costs manufacturers between $100 to $175 per vehicle. As an incentive to assume this extra cost, manufacturers receive a special fuel economy calculation that enables manufacturers to boost their fleet CAFE by up to 1.2 mpg toward complying with CAFE standards. This means that

producing flex-fuel vehicles and obtaining the benefit of the special fuel economy calculation has the effect of allowing manufacturers to comply with a lower CAFE standard than they otherwise would be required to meet. As a result, the Dual Fuel program has weakened the CAFE program's effectiveness in reducing oil consumption in the short-term, both because it lowers the fuel economy standards with which manufacturers must comply and because most flex-fuel vehicles are usually run on regular gasoline.[30] Furthermore, manufacturers have generally put flex-fuel capacity in their larger, relatively lower fuel economy models, particularly light trucks. For example, about 80 percent of flex-fuel vehicles available in model years 2006 and 2007 were light trucks. Light trucks in general must meet a lower CAFE standard than cars, and represent about 50 percent of the new car market. That manufacturers can build these vehicles to an even lower fuel economy standard if they produce light trucks with flex-fuel capabilities, and because these vehicles usually run on gasoline, this erodes potential reductions in fuel consumption that could otherwise come from the CAFE program. Also, as discussed later in this report, our previous work found it is not clear whether the Dual Fuel program has actually increased the availability or use of alternative fuels like E85. For example, although the number of fuel stations offering E85 has increased since 2004, fewer than 1 percent of fuel stations in the country offered E85 as of early 2007.

Some of the weaknesses that we identify here, such as the potential negative safety impact from raising current car CAFE standards and the distinctions the program makes between foreign and domestic cars could be remedied through revisions to the car CAFE program. However, NHTSA does not have the authority to make changes to the car CAFE program, though the administration has requested this authority from Congress, and a bill the Senate passed in the 110th Congress would give NHTSA that authority.[31]

EXPERTS SUGGEST THAT REFINEMENTS TO THE CAR CAFE PROGRAM COULD INCREASE FUEL SAVINGS AND ADDRESS SOME PROGRAM WEAKNESSES

Experts with whom we spoke suggested that several refinements to the structure of the program could increase fuel savings and address weaknesses in the program. The refinements selected for discussion represent those supported by many of these experts and, in some cases, were also supported by research. In addition, we included refinements based on our work on 21st Century Challenges, which concluded that a fundamental review of major program and policy areas can serve the vital function of updating these programs to meet current and future challenges.[32] This is especially important for programs and policies designed decades ago to respond to trends and challenges that existed at the time of their creation. While these refinements show promise to enhance the CAFE program, additional analysis of the potential outcomes would be needed before implementation. Proposed refinements to the CAFE program include the following:

- *Reform the car program to an attribute-based system, as NHTSA recently reformed the light truck program.* In changing the light truck system to a footprint-based approach, NHTSA cited several benefits, including increased fuel savings, enhanced safety, and a more equitable framework for manufacturers because compliance costs are spread more evenly across the industry. Experts with whom we spoke generally agreed with NHTSA that these changes enhanced the light truck CAFE program. NHTSA has requested authority to convert the car program to an attribute-based system, and anticipates that it would use size as it has for the light truck program but has indicated that it might perform some research to confirm size is the best attribute.
- *Periodically review the basic structure of the CAFE program.* A regular and periodic review of the basic structure of the CAFE program could allow NHTSA to ensure that the program keeps pace with current conditions like changes in the fleet mix so that the program's effectiveness in producing oil savings could be maximized, assuming Congress grants NHTSA the authority to make changes to the program's structure. Such a review could also be used to determine whether its new size-

based system for light trucks is increasing fuel economy as intended.

- *Remove incentive for manufacturers to classify cars as light trucks*: Currently, the definitions of cars and light trucks are structured in a manner that allows manufacturers to make modest design changes in order to classify a vehicle as a light truck, and thus meet a lower CAFE standard. For example, vehicles capable of off-highway operation (i.e., four-wheel drive) or that have removable seats to expand cargo space may be considered light trucks. However, recent changes in fleet mix and the use of light trucks (i.e., primarily as passenger vehicles), for example, make the definition outdated. NHTSA recently took some steps to address this concern by issuing revised criteria for classifying vehicles as light trucks, including requiring a vehicle to have three rows of seating to qualify as a light truck. This will result in the removal of wagon-type vehicles such as the PT Cruiser from the light truck classification. However, this concern could be further addressed by an additional revision to the definition of light trucks that more accurately captures attributes of vehicles used for light duty work. Alternatively, if NHTSA implemented an attribute-based system for cars, the distinction between cars and light trucks could be eliminated, and fuel economy standards could be based on attributes.
- *Allow CAFE credit trading between vehicle fleets and among manufacturers*: As discussed previously, if manufacturers exceed the required fuel economy in a certain year, they earn credits that can be applied to past or future model year fuel economy numbers. These credits cannot be traded among manufacturers or between fleets (that is, between cars and light trucks). Greater flexibility in the use of CAFE credits—specifically, trading among manufacturers as well as transferring between fleets—than is now afforded could reduce compliance costs to manufacturers. Specifically, manufacturers for whom it would be particularly costly to achieve a CAFE standard for a particular fleet could trade with another manufacturer who could achieve the standard at less cost or transfer credits between the car and light truck fleets or their foreign and domestic car fleets. Although credit trading would give manufacturers flexibility in how they meet CAFE

standards, the fleet would still need to meet the overall standard. For example, if one manufacturer exceeded the car CAFE standard under the current system by 1 mpg, it could sell that 1 mpg credit to a manufacturer that was 1 mpg under compliance. Collectively, the average of both manufacturers would meet the CAFE standard. In the 2007 *State of the Union* address, President Bush proposed a credit trading system under which manufacturers could trade CAFE credits with one another to improve fuel economy at the lowest possible cost and the Senate passed a bill in June 2007 that would institute a program where manufacturers could trade accrued CAFE credits with one another. As of July 2007, the House has not acted on this bill. Industry representatives have indicated that they would not trade credits with other manufacturers due to competitive concerns, but they thought that many manufacturers would trade within their own fleets, such as between their car and light truck fleets, if that option was available.

- *Raise CAFE penalties with inflation*: CAFE penalties for noncompliance were established as a part of the program and were first applied to model year 1983. NHTSA increased the penalty in 1997 from $5.00 to $5.50 per 0.1 mpg below the standard per vehicle, but it has not increased them since then. Most manufacturers—including all domestic manufacturers—comply with CAFE and do not pay penalties, and it is not clear whether an inflation-based increase in penalties would cause noncompliant manufacturers to comply. However, in previous work, we have recommended that agencies collecting penalties should review their programs regularly to determine if penalties need to increase to ensure that they continue to deter noncompliance.[33] Because CAFE penalties have not risen since 1997, despite increases in inflation, noncompliance now costs less, in real terms, for manufacturers than it did before 1997. If CAFE penalties had kept pace with inflation since NHTSA raised the penalties in 1997, they would currently be set at around $7 per 0.1 mpg for 2007.

- *Eliminate or revise the dual fuel credit*: As previously noted, the Dual Fuel program has the effect of allowing manufacturers to meet lower CAFE standards, and it is not clear to what extent the program has helped increase the production and availability of alternative fuels. Of those who commented, many experts with whom we spoke thought this program should be eliminated or at least revised. For example, the credit could be granted for flex-fuel vehicles sold in states that have a higher concentration of fueling stations offering E85. Alternatively, a lower CAFE credit than the maximum 1.2 mpg credit currently available could be provided. Given that flex-fuel vehicles are not always run on alternative fuels, lowering the credit to more accurately reflect how often these vehicles are actually run on alternative fuels could be appropriate.

The Senate recently passed legislation that would make several changes to the CAFE program, including revising the car CAFE program to an attribute-based program and allowing manufacturers to trade with each other CAFE credits they accrue for exceeding the standards.

NHTSA GENERALLY HAS THE CAPABILITIES TO REFORM CAFE STANDARDS AND ACT QUICKLY IN THE FUTURE, BUT SOME CAPABILITIES COULD BE IMPROVED

NHTSA's recent reform of the light truck CAFE program showed that the agency generally has the capabilities to reform standards and could act quickly in the future to reform the car program, but some of NHTSA's capabilities could be improved. To reform the light truck program, NHTSA leveraged the work of outside experts. For example, in 2001, at the direction of Congress, NHTSA contracted with the National Academies of Science to conduct a peer-reviewed study of CAFE and automotive technologies. The NAS report included several findings and recommendations and a study on the feasibility of automotive technologies for increasing fuel economy in the future. The study, completed in 5 months, was the basis for much of NHTSA's rule-making affecting light trucks produced in model years 2008 through 2011.

To solicit additional input and ensure openness in its deliberations, NHTSA published advance notices to collect information from the automotive community and others with expertise in CAFE to assist in developing a proposed light truck rule. NHTSA received over 45,000 comments, and NHTSA officials stated that they changed the final rule to use size instead of weight as the attribute on which CAFE standards would be based and revised some of its assumptions in producing the final rule for the light truck rule, based on information provided in the comments. For example, NHTSA officials stated that they revised their analysis and assumptions related to the rate at which it was practicable for manufacturers to add fuel-saving technologies to their fleet.

In developing the revised light truck CAFE standard, NHTSA also used a computer model to help estimate the costs and benefits of increasing CAFE standards. NHTSA worked with DOT's Volpe National Transportation Systems Center to develop the model. Also, because of its past work with the automotive industry producing previous light truck standards, NHTSA has established a good working relationship with the automotive industry. Officials at one automotive organization said NHTSA properly handled its confidential data and produced science-based results.

While, in general, NHTSA had the capability to reform the light truck program in a manner supported by the automotive experts, manufacturers and safety experts with whom we spoke, these stakeholders said that there are areas where NHTSA could improve its capabilities for managing and revising the CAFE program in the future. For example, some experts observed that NHTSA has lost staff since the 1990s and stated that this reduction may stem from the congressional prohibition on NHTSA's making any changes to CAFE. NHTSA officials told us they need an additional staff member with expertise in automotive engineering and computer modeling to assist NHTSA in estimating the potential impact of new technologies on fuel economy and to perform other tasks in preparation for possible future changes to CAFE standards. Also, NHTSA currently relies on the Volpe Center and the NAS report to provide the detailed information on the capabilities of new technologies that NHTSA uses to set future CAFE standards. Such independent information is important to NHTSA when developing CAFE standards. However, NHTSA officials told us that they rely heavily on the technological assumptions related to the impact of new technologies on fuel economy in the 2002 NAS report and that they fear the study's assumptions are becoming out-of-date. These officials stated they would like to update the NAS study and have requested additional staff and funding for an update of the NAS study in NHTSA's fiscal year 2008 budget request.

Lastly, several stakeholders and experts said they were concerned about certain inputs that NHTSA officials used in the computer model maintained by DOT's Volpe Center. NHTSA uses this model as a tool to help estimate the fuel savings that will result from CAFE increases and to estimate is the likelihood that manufacturers will comply with future CAFE standards, based on the confidential data NHTSA received from the manufacturers. Specifically, some experts were critical because NHTSA and Volpe staff did not assign a dollar value to reductions in greenhouse gas emissions that would result from an increased standard. NHTSA officials said they did not assign a value because the scientific community had not reached a consensus on the worth of reductions in carbon dioxide emissions, though researchers have developed a range of values that could be considered. Therefore, according to one expert, the results of the model may underestimate the total dollar benefits to society of raising CAFE standards, since the dollar value of reduced greenhouse gas emissions was not included in the model's results. Revisions to the car CAFE program, if they occur, may provide an opportunity to revisit this issue and to conduct additional sensitivity analyses, possibly in conjunction with other government agencies such as DOE and EPA, to examine how alternative values for greenhouse gas emission reductions affect the model's results. NHTSA has indicated it will examine this issue in the next CAFE rule making.

Chapter 4

SOME MARKET-BASED POLICY OPTIONS COULD COMPLEMENT THE CAFE PROGRAM

Through reviews of our past reports and other studies, interviews with experts, reviews of recently proposed legislation, and analysis of existing programs in the United States and other countries, we identified several market-based policies involving cars and light trucks that could complement and strengthen CAFE's fuel-saving effects or that could be broader reaching and potentially more cost-effective alternatives to the CAFE program. The policies discussed in this section represent those that experts viewed as most promising to reduce fuel consumption by cars and light trucks. Market-based consumer incentives could complement CAFE by increasing consumer interest in purchasing fuel-efficient vehicles, and some incentives already exist. However, some of these incentives may work at cross purposes to programs intended to reduce fuel consumption. Also, although some policies we identified could complement CAFE's fuel-saving effects, the policies may not be able to produce large enough fuel savings to achieve broader goals in the future. Market-based incentives have also been used to increase the availability and use of biofuels, but our recent report on these efforts identified several limitations, and the cost-effectiveness of these programs is unclear.[34] Several options, including a tax on fuel or a carbon cap-and-trade program, affect a broader range of fuel-saving behaviors among consumers and could be more cost-effective. Such options could help the nation reach larger, long-term fuel-saving goals at a lower cost than CAFE, but time would be needed to design and garner support for each before it could be implemented.

MARKET-BASED CONSUMER INCENTIVES FOR PURCHASING FUEL-EFFICIENT VEHICLES EXIST, BUT THEY ARE NARROWLY TARGETED AND HAVE IMPLICATIONS FOR FEDERAL SPENDING

Market-based incentives to encourage consumers to choose higher fuel economy vehicles may be particularly important as options to complement CAFE. Specifically, while CAFE encourages a supply of vehicles with a relatively high fuel economy, it does not create a demand for them. Auto manufacturers with whom we spoke told us that consumers generally choose a vehicle based on other attributes, such as performance, interior and trunk capacity, and safety features, though recent high gasoline prices have had some impact on the demand for higher fuel economy. Consumer incentives could help create a stronger market for vehicles with higher fuel economy, which could encourage manufacturers to develop new fuel-saving technologies more quickly. A few policies that encourage a market for fuel-saving vehicles are currently in place, and while we identified weaknesses with existing incentives, such policies could be improved to complement any efforts Congress takes to improve the CAFE program. These policies are described in the following sections.

Tax Credits Can Encourage Consumers to Purchase Vehicles with a Higher Fuel Economy, but Related Costs Must Be Considered in Designing Such a Policy

The Energy Policy Act of 2005 established a tax credit for the purchase of a hybrid vehicle, which is propelled by a standard gasoline (or diesel) internal combustion engine in combination with an electric motor and battery storage system.[35] Hybrid technology can significantly improve fuel economy—for example, according to the DOE's *Fuel Economy Guide*, the most efficient model year 2007 hybrid car is rated at 60 mpg for city driving and 51 mpg on the highway. The tax credits range from $250 to $3,400, depending on the fuel economy of the model; and the credit is phased out once a manufacturer has sold 60,000 vehicles. The 60,000 vehicle limit was intended to prevent tax credits from accruing excessively to foreign hybrid manufacturers. Almost 216,000 model year 2006 hybrids have been sold.

Although recent surges in gasoline prices above $3 per gallon may be changing consumer behavior, previous research has found that consumers purchasing new vehicles consider several factors in choosing a model, but fuel economy has not typically been a priority. Of those experts who discussed the issue with us, most supported the use of tax credits to encourage consumers to place a higher value on fuel economy. A recent report by the Center for Clean Air Policy[36] noted that credits can lower the cost of a fuel-saving car, thus making these vehicles more appealing to consumers, and also can encourage manufacturers to roll out new technologies in their fleet by helping to overcome market barriers. Specifically, cars with new technologies are generally more expensive than those with conventional technologies because it takes time for manufacturers to reach economies of scale, and some portion of these costs are passed onto the consumer. Tax credits can help to offset the cost differential between cars with advanced and conventional technologies, which means that consumers will not face as much of a price disincentive for choosing a car with new fuel-saving technologies.

One weakness of the hybrid tax credit that some experts identified is that by targeting specific technologies, such credits may give an advantage to technologies that ultimately are not the most efficient or cost-effective technology available to achieve fuel-saving goals. For example, the current tax credits that encourage consumers to purchase vehicles with hybrid technology may discourage the development of other promising fuel-saving technologies, because those technologies would not have the cost advantage of a tax credit to support their sale.

To address this weakness, some experts suggested offering tax credits based on a performance standard. For instance, a credit could be provided for any vehicle achieving a fuel economy higher than 40 mpg, regardless of the technology the vehicle uses. Such an approach could also support environmental goals by including performance measures related to pollution emissions as well. This approach would target a broader range of fuel-saving technologies but could also increase the costs of the policy to the federal government. As we have stated in recent reports, tax credits are a type of tax expenditure that results in revenue loss for the federal government, and as such, they need to be evaluated to determine if their benefits in achieving clear, outcome-oriented goals exceed their costs.[37]

One option that would address the costs associated with providing credits for purchasing vehicles with a higher fuel economy is a feebate program, which would incorporate both incentives and disincentives by taxing the purchase of vehicles that achieve a lower fuel economy and applying those revenues to subsidize a rebate or credit for the purchase of vehicles that achieve a higher fuel economy. Although the amount of the fees and rebates might need to be relatively high to affect consumer choices,[38] the system could be designed to be revenue-neutral, where the amount of rebates paid out is covered by the fees collected. In addition, feebates can be adjusted as CAFE standards are increased to ensure that there is always a market element to complement CAFE. Such a system is being considered in Canada to complement Canada's voluntary fuel economy standards.

One limitation noted by some of the experts with whom we spoke—and a potential reason to use feebates to complement rather than replace CAFE—is that feebates have not been tested on a large scale, and the market may not respond as expected. In addition, some industry representatives told us that such a system should be national, rather than state-initiated, to prevent car buyers from going to certain states to buy vehicles that achieve higher fuel economy so they can obtain a rebate or, conversely, going to other states to buy vehicles that achieve lower fuel economy to avoid paying a fee.

Taxes Can Discourage the Purchase of Vehicles with a Low Fuel Economy and Provide a Revenue Stream for Other Fuel-Saving Programs, but Can Face Consumer Resistance

Taxes on vehicles with a low fuel economy are another type of market-based incentive to encourage consumers to choose vehicles with a higher fuel economy and are another option to complement the CAFE program. Such taxes have been implemented in the United States and other countries. Specifically, consumers can buy a vehicle with a high fuel economy without paying a tax penalty or buy a less fuel-efficient vehicle that fits other needs, but they will incur a tax penalty. The public benefits from either consumer decision, through fuel savings or collection of revenue that the government can put toward other fuel-saving programs—for instance, federal research and development programs on fuel-efficient technology or alternative fuels.

The U.S. Gas Guzzler Tax is an example of an existing disincentive against purchasing vehicles that obtain relatively low fuel economy ratings. The tax is levied on the sale of new cars whose fuel economy does not meet certain levels. The tax is paid by the manufacturer, which must disclose the amount to potential buyers by including it on the fuel economy window sticker. The tax applies only to cars and not to light trucks, and the tax is collected by the Internal Revenue Service. Manufacturers currently begin paying a tax when their cars obtain less than 22.5 mpg, and the tax increases incrementally for cars with lower fuel economy (see fig. 4). In general, manufacturers of luxury or sports cars primarily pay the Gas Guzzler Tax, such as Aston Martin, Ferrari, and Mercedes.

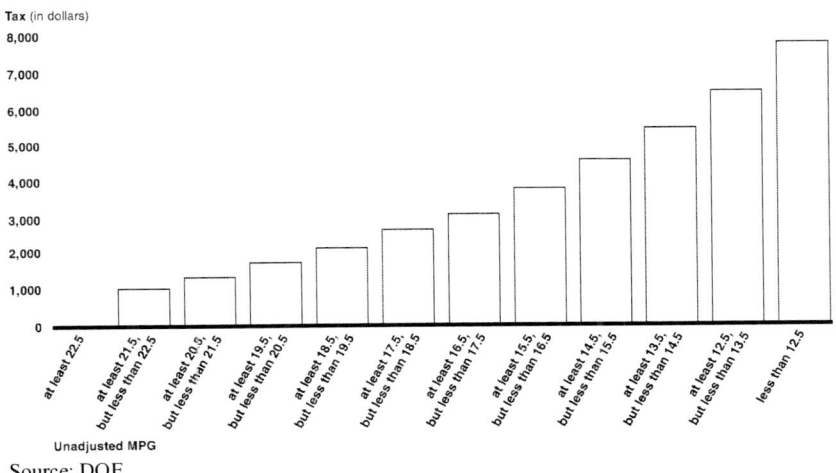

Figure 4. Gas Guzzler Tax Structure.

Several issues may limit the effectiveness of the Gas Guzzler Tax. First, although the tax was intended to discourage the production and purchase of vehicles obtaining a low fuel economy, its structure has not been updated since 1990, and the extent to which the tax serves as an effective disincentive is not clear. Because the amount of the tax has not been adjusted for inflation since 1990, it is less expensive for manufacturers to pay the tax now than it was in years prior to 1990, so the tax might be less of a disincentive now than in the past. Second, as previously noted, light trucks are not subject to the Gas Guzzler Tax. In 1979, the year before the Gas Guzzler Tax took effect, light trucks accounted for about 10 percent of the new light vehicle market. By 2004, light trucks accounted for almost 53 percent of the new light vehicle market and, according to NHTSA, in many cases are primarily used as passenger vehicles, despite having low fuel economy. This is a significant change in the conditions of the auto market, one that the original lawmakers who developed the tax may not have anticipated. Finally, it is not clear to what extent the Gas Guzzler Tax encourages consumers to choose a vehicle with a higher fuel economy. As noted, the tax generally is paid by manufacturers of luxury and sports cars. If the tax were applied to a broader range of vehicles—for example, by increasing the fuel economy standard to which the tax applied—the tax could influence more consumers' car purchasing decisions. While expanding the Gas Guzzler Tax would encourage consumers to buy fewer vehicles subject to this tax, those already owning such vehicles before the tax goes into effect may choose to hold onto those vehicles longer than they otherwise would. If new cars subject to an expanded Gas Guzzler Tax had better fuel economy than these cars, then holding onto them longer would be at cross purposes with the objective of reducing fuel consumption.

One alternative to the Gas Guzzler Tax that has been implemented in other countries is a structure of graduated registration fees that corresponds to different levels of fuel economy. This type of tax is paid yearly with the renewal of an owner's vehicle registration rather than only once at the time of purchasing a new car, and the fee increases as a vehicle's fuel economy rating decreases. Denmark, France, and the United Kingdom have implemented a graduated registration tax, and the cost of registering a fuel inefficient vehicle can be high. For example, if the current structure of Denmark's "Green Owner Tax" were applied in the United States, it would cost annually as little as about $30 to register a fuel-saving compact car, compared with about $1,160 for a luxury sedan with a much lower fuel economy.[39] Such recurring fees may increase the value consumers place on purchasing vehicles with higher fuel economy ratings. In addition, while the Hybrid Tax Credit and the Gas Guzzler Tax apply only to new model year cars being sold, graduated registration fees would apply to all vehicles, and therefore might influence consumer choices, even for used vehicle purchases. Furthermore, because graduated registration fees would apply to all vehicles, they could have a less adverse effect on the market for new cars than a tax on new cars only.

Other Tax Policies with Different Goals Might Affect Consumer Choice of Vehicles

Other tax incentives that are designed to support goals other than reducing oil consumption, but that nonetheless affect consumer choices in purchasing a vehicle, may negate some benefits from oil-saving programs. For example, small businesses can obtain a tax savings through depreciation write-offs for the purchase of an SUV over 6,000 lbs. The depreciation write-off on cars, including hybrids, are treated less generously, offering much smaller write-offs due to more stringent depreciation limitations As a result, businesses seeking to maximize a tax write-off may choose to purchase an SUV, which generally have lower fuel economy ratings than hybrid cars. In addition, tax laws such as those that exclude from income and payroll tax a portion of employer-paid parking expenses may encourage individuals to commute by passenger car or light truck, which could increase fuel consumption. Although these laws were not intended to save fuel, the majority of experts with whom we spoke thought that policy should be integrated and aligned to produce fuel savings. In addition, we have recommended that government programs be periodically reexamined to ensure that they are meeting current challenges and national goals.[40]

TAXES ON GASOLINE, CARBON EMISSIONS, OR VEHICLE MILES TRAVELED COULD AFFECT A BROADER RANGE OF CONSUMER DECISIONS THAT RELATE TO FUEL CONSUMPTION

Other tax options, including a tax on gasoline or carbon emissions[41] would create incentives that could affect a broader range of consumer choices, including how much to drive, whether to use vehicles with a higher fuel economy, and when to retire older, less efficient vehicles. A tax on the number of miles driven by an individual (vehicle miles traveled tax or VMT tax) would encourage consumers to drive less. However, unlike a gasoline or carbon tax, a VMT tax does not vary depending on how many mpg a vehicle achieves; thus, it does not provide a direct incentive to purchase a vehicle with a higher fuel economy. Because a gasoline or carbon tax could have such broad effect on consumer decisions, it could be used to complement CAFE or, if set at an appropriate level, to replace CAFE standards. The economic literature we reviewed indicates that a gasoline or carbon tax would produce greater oil savings than increasing CAFE standards alone and at less cost. Furthermore, this literature and all of the economists with whom we spoke stated that a tax on gasoline or carbon would be cost-effective, whereas increasing CAFE standards would not be as cost-effective. For example, CBO estimated that increasing the gasoline tax to achieve a 10 percent reduction in fuel consumption would cost far less than an increase in CAFE standards.[42]

In addition to being cost-effective and influencing a broader range of consumer decisions than tax incentives on new car purchases, a gasoline or carbon tax offers a number of other benefits in terms of potentially reducing fuel consumption:

- It would result in a wide range of fuel-saving responses from all consumers rather than only from those purchasing a new vehicle. For example, a higher tax on gasoline or carbon would provide a financial incentive for all drivers to buy vehicles with higher fuel economy, retire vehicles with lower fuel economy sooner, and drive less. By comparison, CAFE standards or consumer incentives to buy vehicles with a higher fuel economy influence a much smaller group of consumers—namely, those choosing to purchase a new vehicle, which limits the effects of these strategies on fuel consumption. In addition, because increases to CAFE can increase the cost of new vehicles through the addition of new technology, CAFE can slow the sale of new cars and extend the life of older vehicles, which may have lower fuel economy ratings.
- Higher gasoline prices resulting from either a gasoline or carbon tax could sustain consumers' interest in fuel-saving vehicles, leading to a more predictable demand for these vehicles, which is important to the car manufacturing industry. Industry representatives told us that it is difficult for them to respond to rapid changes in consumer interest triggered by fluctuations in fuel prices because auto manufacturers generally plan their products years in advance. For example, in 2005 Hurricane Katrina and other factors caused disturbances in regional gasoline supplies, and gasoline prices climbed to a nationwide average of almost $3 per gallon. During this time, sales of light trucks declined, causing manufacturers like Ford to significantly reduce production.
- A gasoline or carbon tax could complement increased CAFE standards by helping address the rebound effect—an increase in driving among those with fuel-saving cars because the per-mile cost of driving is lower. The rebound effect reduces the fuel savings that can be produced by increasing CAFE standards.[43] A gasoline or carbon tax would provide a financial incentive for consumers to drive less, which could mitigate the rebound effect.

- We recently reported that additional taxes on oil or carbon would be the most economically efficient means of increasing the production and use of biofuels because those taxes would allow biofuels to be used at the level where they provide the greatest economic, environmental, and other benefits.[44]
- Some revenues from the gasoline and carbon tax could be "recaptured," or used to fund other efforts to reduce fuel consumption, such as funding research and development of fuel-saving technologies for cars and light trucks. The current federal gasoline tax is $0.184 per gallon, of which $0.183 goes to fund highway and mass transit trust funds.

An alternative to a gasoline or carbon tax that more directly addresses the effect of increased driving on oil consumption is a VMT tax. The number of overall vehicle miles traveled has increased by 22 percent from 1994 to 2003, and increases in VMT result in increased fuel consumption, pollutants and carbon emissions, congestion (which further increases fuel consumption), and road maintenance requirements. A VMT tax effects drivers' choices about how much to drive, and therefore, could help the nation meet several goals. Also, it could be used to complement CAFE standards and could address the rebound effect by creating a disincentive for people to drive more when improved fuel economy makes driving less costly.

In 2006, Oregon tested the feasibility of replacing the state gasoline tax with a VMT tax. The Global Positioning System (GPS) was used to track the miles driven, and participants pay the VMT tax ($0.012 per mile traveled) instead of the state gasoline tax when they fill up at gasoline pumps that can read information from the GPS. Using a GPS could also track mileage in high congestion zones, and the tax could be adjusted upward for miles driven in these areas or during more congested times of day such as rush hour—a strategy that might reduce congestion and save fuel. In addition, the system could be designed to apply different tax levels to vehicles, depending on their fuel economy. On the federal level, a VMT tax could be based on odometer readings, which would likely be a simpler and less costly way to implement such a program.

Some limitations exist for a gasoline, carbon, or VMT tax. For example, the effectiveness of such taxes in reducing fuel consumption would depend in part on setting the tax at a level that would change consumer behavior. In addition, each of these taxes would increase the overall costs of driving, which could disproportionately affect rural residents, who often must drive more because of limited public transportation and greater distances to obtain services, and low-income drivers. Some economists believe that this disadvantage can be addressed through "revenue recycling," a measure in which behaviors considered to be valuable to the economy are lowered to offset some or all of an increased tax on behaviors that create additional costs for the public. For example, taxes on income could be lowered to offset increased taxes on gasoline consumption or miles driven. In addition, a VMT tax—unless it is adjusted based on the fuel economy of the vehicle—does not provide incentives for customers to buy vehicles with higher fuel economy ratings because the tax depends only on mileage. Also, because the tax would likely be collected from individual drivers, a VMT tax could be expensive for the government to implement, potentially making it a less cost-effective approach than a gasoline or carbon tax. By comparison, the government collects the federal gasoline tax from fuel producers, not individual consumers, which simplifies and lowers the cost of administering the tax. However, the most difficult obstacle for the use of a gasoline, carbon, or VMT tax continues to be public resistance, which stems from the high visibility to the consumer of the cost of these types of taxes. By comparison, policies like CAFE also create costs to the consumer—such as a higher price for new vehicles due to new technology to save fuel—but these costs may be less obvious to consumers because they are incorporated in the sale price of the vehicle.

EFFORTS TO EXPAND THE MARKET DEMAND FOR BIOFUELS HAVE BEEN INITIATED, BUT SEVERAL BARRIERS IMPEDE PROGRESS

Developing renewable and alternative fuels has been a prominent part of the administration's plans to reduce oil consumption. For example, in the *State of the Union* address in January 2007, the President established a goal to reduce gasoline consumption by 20 percent of projected use in 2017. About 15 percent of oil savings will come from renewable and alternative fuels and 5 percent is expected to come from increased CAFE standards. Fuels such as ethanol—which is made from renewable feedstocks like corn—are currently available on the market, while other renewable fuels, like cellulosic ethanol—which is made from sources like corn stalks that are in abundant supply—shows promise although technological advances are still needed to reduce the cost of its production. Biofuels offer several environmental advantages compared with other types of alternative fuel, including coming from renewable resources and emitting lower levels of carbon dioxide when they are consumed compared with conventional gasoline and alternative fuels such as those produced by converting coal to liquid fuel.

Expanding the use of alternative fuels can work in parallel with CAFE standards to reduce oil consumption. Although fuel economy standards do not create an incentive for consumers to seek opportunities to use biofuels, as we reported in June 2007, strategies to develop both the supply and demand for biofuels in the transportation sector are currently in place, but several barriers impede progress.[45] We found that although the production of ethanol, one of the most commonly available biofuels for cars and light trucks, has increased significantly,[46] most of this supply is being used as an additive in gasoline (10 percent or less) to improve the emissions of conventional gasoline and to extend gasoline supplies rather than being made into the alternative fuel, E85. In addition, few fueling stations offer E85—in early 2007 approximately 1,100, or fewer than 1 percent of the fuel stations in the United States, offered E85, and these were primarily concentrated in the Midwest. As our June 2007 report indicated, other significant barriers to expanding the availability of E85 also exist, including higher costs of production, limits on available land to grow the feedstocks used to create E85, and increases in food costs associated with greater use of corn and soybeans to make these fuels instead of food products, which may discourage use of biofuels.

To support public and private investments in expanding the production and availability of alternative fuels, including biofuels, two programs work to expand the market demand for these fuels: (1) an incentive that aids efforts to meet the CAFE standards for auto manufacturers and (2) requirements that federal agencies purchase flex-fuel vehicles. As noted earlier in the report, auto manufacturers receive a maximum increase of 1.2 mpg toward meeting CAFE standards for producing flex-fuel vehicles capable of running on E85 or conventional gasoline. Although manufacturers have increased their production of flex-fuel vehicles and more models are available now than in the past, a 2002 DOT, EPA, and DOE report estimated that less than 1 percent of the fuel consumed by these vehicles was E85, though this number may be higher now that E85 is in greater supply. As noted, E85 is not widely available to consumers and while most E85 fueling stations are located in the Midwest, we recently reported that, according to the Alliance of Automobile Manufacturers, in 2006, the largest number of privately owned flex-fuel vehicles were in Texas, Florida, and California.

A second program to increase the availability of flex-fuel vehicles and alternative fuels by increasing demand for both was included in the Energy Policy Acts of 1992 and 2005, which required federal agencies to increase their purchase of flex-fuel vehicles and use alternative fuels to fuel these vehicles. We recently evaluated the extent to which the United States Postal Service (USPS) has been able to comply with these requirements, and we reported that these requirements may not be contributing to passenger vehicle oil savings. For example, to comply with the laws, USPS purchased a large fleet of flex-fuel vehicles to reduce its reliance on petroleum-based fuels, yet the limited nationwide alternative fueling infrastructure and the often higher cost and lower efficiency of E85, compared with regular gasoline have prevented USPS from using alternative fuels. As of 2006, alternative fuels accounted for only 1.5 percent of the total fuel consumed by USPS's internal fleet. USPS officials have had success in improving gasoline mileage by using hybrids, which the officials indicated are well suited to the stop-and-go driving of mail delivery, but hybrids are not considered flex-fuel vehicles and therefore do not help USPS comply with the Energy Policy Act of 2005.[47]

The federal government also uses tax credits to promote the greater availability and use of biofuels. For example, the American Jobs Creation Act of 2004 created a tax credit for ethanol use that provides a 51 cent per-gallon tax credit to fuel blenders for ethanol they blend with gasoline as well as tax credits for installing fuel stations providing alternative fuels.[48] We recently reported that, according to Treasury Department data, these credits cost the government about $2.7 billion in forgone revenue in 2006. Whether these credits create energy independence or environmental benefits sufficient to justify their costs is a matter of debate.[49]

A CARBON CAP-AND-TRADE PROGRAM WOULD COMBINE REGULATORY AND MARKET-BASED ELEMENTS, BUT INCLUDING CARS AND LIGHT TRUCKS COULD BE COMPLICATED

Several bills have been introduced in both the House and Senate proposing a multi-industry cap-and-trade program to reduce greenhouse gas emissions, including carbon dioxide emissions. Cap-and-trade programs combine a regulatory limit or cap on the amount of a substance—in this case, carbon dioxide—that can be emitted into the atmosphere with market elements like credit trading to give industries flexibility in meeting the cap.[50] The cap can be reduced in outlying years in order to steadily decrease the total amount of carbon dioxide emitted; and, in this scenario, individual companies would comply with the cap by either reducing their emissions to the cap's limit or buying credits from a company that is below the cap. Because burning gasoline produces carbon dioxide emissions, a cap on carbon dioxide, if applied to cars and light trucks, would also improve fuel economy and reduce fuel consumption.[51]

Research indicates that by combining regulatory (namely the carbon cap) and market-based elements (such as credit trading), cap-and-trade programs can produce cost-effective outcomes, especially when compared with regulatory programs. For example, the cap sets a predetermined limit on emissions, but credit trading allows the industry to achieve the goal in the least costly manner by allowing companies for whom compliance costs are low to overcomply and sell allowances to those companies for whom compliance costs are high, all while remaining within the overall cap. In addition, the costs are borne and shared by all those industries participating in the program—and some portion of these compliance costs are passed on to the consumer. Research also suggests that for a carbon cap-and-trade program to maximize its cost-effectiveness, it would need to include all major sources of carbon emissions from a broad range of industries, such as electric utility companies, oil producers, auto makers, and others, which would spread the cost of compliance broadly.[52]

Designing a cap-and-trade program would be complicated and would take time to develop, and its effectiveness in producing fuel savings and reduced greenhouse gas emissions would depend in part on how aggressively the cap was reduced. For example, when a program is established, the government must give or auction allowances for the right to emit carbon dioxide up to the total number of allowances equal to the cap. Determining whether to auction or give allowances to companies is important in designing a cap-and-trade program because it has cost implications for the government and society and can create competitive advantages for participating companies. For example, if allowances are auctioned, the government will receive revenues, which could be used to offset the costs of managing the program or fund research and development on technology to reduce carbon emissions.

In addition, a carbon cap-and-trade program could be designed to incorporate cars and light trucks, which would influence fuel consumption but would also create additional design challenges that would impose different requirements and costs on auto manufacturers. One approach would introduce an "upstream" cap on fossil fuels in which all producers and importers of oil, coal, and other fossil fuels would be required to hold allowances based on the carbon content of their fuel. Such a design would link the pricing of transportation fuels to their carbon content, which would in turn affect consumer behavior in a similar manner to a carbon tax by encouraging consumers to buy more fuel-saving vehicles and drive less. This approach would not require a CAFE standard or any type of carbon cap on auto manufacturers, but instead it would allow fuel pricing to drive changes in the market.

Alternatively, some proposals under consideration in Congress would establish a cap-and-trade program and would include some form of cap for auto makers. This could be accomplished by using CAFE standards as a proxy for carbon emissions, increasing the CAFE standards over time, and developing a credit trading program between CAFE credits and the carbon trading among other industries. However, maintaining the CAFE system within a larger cap-and-trade program could result in higher compliance costs for auto manufacturers, making it more costly for manufacturers to reduce emissions, compared with other industries.

Chapter 5

CONCLUSIONS

Reducing the nation's growing oil consumption, particularly for cars and light trucks, is a formidable challenge. Despite its limited scope, the CAFE program has reduced oil consumption by cars and light trucks over what it would have otherwise been, and the evidence suggests that increasing CAFE standards would save additional oil. However, the average vehicle fuel economy of cars and light trucks in the United States has stagnated since 1990 due to several factors, including the low price of oil during much of this period and an increase in the number of large cars and SUVs in the market for which there have not been comparable increases in CAFE standards. Most, but not all, manufacturers have been exceeding the car CAFE standard for some time and therefore do not oppose incremental increases in these standards. Furthermore, experts with whom we spoke, and NAS in its 2002 report, stated that the technology exists to increase fuel economy without large increases in vehicle costs.

As shown by its recent revision of the light truck CAFE program, NHTSA has the technological capabilities to perform the analysis required to raise the car CAFE standard while balancing fuel economy improvements against concerns about vehicle safety and cost. As a result, NHTSA could move quickly to increase the car CAFE standard and revise the car CAFE program. However, updating the 2002 NAS study would be helpful in giving NHTSA the most up-to-date technological information for determining future CAFE standards. In its fiscal year 2008 budget request, NHTSA has asked for funding to update this study, so that it is not reliant on outdated technological data. Also, under current law, NHTSA does not have the authority it believes it needs to revise the car CAFE program, though the administration has asked Congress several times to provide this authority, without success. Congress could choose to set new standards for CAFE, or it could give NHTSA the authority to reform the car CAFE program, much as it recently revised the light truck program, or both. Either approach would provide an opportunity for NHTSA to evaluate the car CAFE standard and increase fuel economy while attempting to minimize any adverse effects on safety and the equity and consumer choice concerns associated with the current car CAFE program. In addition, evaluating the impact of refinements such as the current CAFE penalty structure and incentives to classify vehicles as light trucks would be an important component to maximize the effectiveness of any CAFE program revisions. The recently passed Senate bill would address some of these refinements, including creating an attribute-based car CAFE program and instituting a system of credit trading for manufacturers.

While CAFE has been an important tool to reduce oil consumption by cars and light trucks and has several strengths, because of its focus on cars and light trucks, the potential oil savings that can be obtained from CAFE may not be enough to help the nation achieve larger fuel-saving goals. Several alternatives to CAFE, including a gasoline or carbon tax or a cap-and-trade program for carbon dioxide, have the potential to produce further fuel savings at less cost and could address a broader range of national goals, including addressing climate change. However, overcoming consumer resistance to a highly visible cost like a gasoline tax or developing a design for a carbon cap-and-trade program that would incorporate cars and light trucks in an equitable and cost-effective manner would both likely require time and consensus-building. In the interim, increases in CAFE standards and revisions to the car CAFE program similar to recent changes to the light truck CAFE program are likely to help the nation make some progress toward reducing fuel consumption. In addition, evaluating and updating existing consumer incentives, such as tax credits for buying fuel-saving vehicles or taxes on purchases of vehicles with low fuel economy ratings, could strengthen the CAFE program's fuel-saving effects. Finally, other market incentives that are designed for other purposes but nonetheless affect passenger vehicle fuel consumption, such as depreciation write-offs for small businesses purchasing large SUVs, also could be evaluated to determine whether the value these programs contribute toward their intended goals is sufficient to offset potential increases in oil consumption.

Chapter 6

MATTERS FOR CONGRESSIONAL CONSIDERATION

If Congress decides to increase CAFE standards, either through setting new standards itself or directing NHTSA to determine the standards, it should consider providing NHTSA with the flexibility and information necessary to reform and revise CAFE standards while mitigating any adverse impact on safety, consumer choice, or competitive equity concerns. Thus, Congress should consider giving NHTSA (1) the authority to reform the car CAFE program much as it restructured the light truck CAFE program and evaluate additional refinements to the program such as credit trading; (2) the resources to update information on the capabilities of new technologies to enhance passenger vehicle fuel economy—as was done in the 2002 NAS study; and (3) the flexibility to adjust the program in the future in response to changes in the passenger vehicle market, such as improved automotive technology and changes in the mix of passenger vehicle types.

Appendix I

SCOPE AND METHODOLOGY

To obtain information on how the Corporate Average Fuel Economy (CAFE) program is designed to reduce oil consumption by cars and light trucks and its status, we reviewed relevant law, including the legislation that established the program and authorized the National Highway Traffic Safety Administration (NHTSA) authority to administer it as well as legislation creating the CAFE credit program for manufacturers of flex-fuel vehicles. We reviewed NHTSA's rule-making documents that reformed the light truck standards, including the advanced notice of proposed rule making, input provided by outside parties during the comment period, and the final rule, paying particular attention to changes between the initial and final rule. We also reviewed program guidance describing the Volpe Center's role assisting NHTSA in setting new CAFE standards as well as material describing Volpe's cost benefit analysis, the variables used in the analysis, and documentation of the rationale for decisions to assign certain values to certain variables. To determine the scope of cars and light trucks that were subject to CAFE, we examined data on new car sales since 1978 and tracked changes in the number of cars and light trucks sold. To further our understanding of how NHTSA works with the Environmental Protection Agency (EPA) to evaluate the fuel economy of new vehicle models, we reviewed relevant legislation and EPA program guidance about CAFE testing fuel economy labeling procedures. We also examined EPA's recent rule to modify the methodology for calculating fuel economy levels posted on new car labels. We reviewed program guidance on NHTSA's process for tracking vehicle model fuel economy and manufacturer credits toward meeting CAFE standards, and we reviewed the process for notifying noncompliant manufacturers in order to understand NHTSA's enforcement procedures. We also examined NHTSA's data on penalty collection since the program's inception. To complement our review of key legislation and program documents, we interviewed a wide range of officials at agencies, including NHTSA, EPA, the Department of Energy (DOE), and the Volpe Center to ensure that we had a clear understanding of the CAFE program's design and implementation.

To obtain information about the strengths and weaknesses of the CAFE program, we interviewed officials from NHTSA, EPA, and DOE, as well as experts in fuel economy and safety who either participated on the 2002 committee for the National Academy of Sciences (NAS) report on CAFE standards or who were recommended by members of the NAS committee or NHTSA. We also interviewed the applicable automobile workers trade union (UAW) and industry groups representing the automobile manufacturers, automotive safety experts, and insurance industry representatives. In addition, we reviewed several key studies, including the 2002 National Academy of Sciences analysis of CAFE and articles by the Congressional Budget Office, our previous work on fuel economy, and other recently published articles about CAFE and its effect on reducing fuel consumption, carbon dioxide emissions, and other benefits. To understand what influence CAFE standards have had on fuel economy, we obtained data on changes in the average fuel economy of cars and light trucks since the initiation of the program. We also examined estimates developed by NHTSA and others about how much additional fuel would have been consumed in the absence of CAFE standards. To ensure that the studies we considered were of sufficient scientific rigor, we limited our review to articles published in well-respected peer reviewed journals and those provided by experts or organizations that we interviewed because of their level of expertise in this area. These articles were reviewed for quality and reliability by our methodologists. To identify potential refinements that could address weaknesses in the CAFE program, we spoke with a wide range of experts and reviewed relevant literature. The refinements selected for discussion represent those supported by many of these experts and in some cases were also supported by research. In addition, we included refinements based on our work on 21st Century Challenges, which concluded that a fundamental review of major program and policy areas can serve the vital function of updating these programs to meet current and future challenges.

To assess NHTSA's capabilities to further revise CAFE standards, we reviewed budgets for the CAFE program and NHTSA's fiscal year 2007 budget. We also reviewed documentation about NHTSA's previous and current staffing levels and plans to hire additional staff. Further, we

Scope and Methodology 53

consulted with experts that were familiar with NHTSA's operation of the CAFE program to discuss whether NHTSA had sufficient staff, whether staff had appropriate technical expertise such as in automotive engineering, and to what extent NHTSA leveraged outside experts from universities, the National Laboratories, and consulting firms. To determine whether NHTSA's use of computer modeling to analyze the costs and benefits of increasing CAFE standards was adequate, we reviewed documentation of the models assumptions, comments submitted during the rule-making process about these assumptions, and we met with NHTSA and Volpe Center staff to discuss the processes and resources they used to assign values to certain variables. We compared this information to guidance published by the Office of Management and Budget for federal agencies using cost benefit analyses to develop policy.

To further our understanding of other market-based policies that are available to replace or complement the CAFE program, we conducted literature searches of recent scholarly publications analyzing options to reduce fuel consumption. We also included in our review any article recommended by the experts with whom we spoke. Our literature review for this section included nearly 100 publications. Finally, to obtain information on other market-based options for reducing oil consumption, we interviewed over 30 experts in fuel economy from universities and advocacy organizations, the National Laboratories, automotive engineering consulting firms, and other industry stakeholders. We selected these experts by contacting officials who served on the 2002 committee for the National Academy of Sciences report on CAFE standards as well as by asking government agencies such as NHTSA, DOE, and EPA to recommend outside experts with whom we should speak. During these conversations, we asked them for names of additional experts we should contact. The experts we interviewed had expertise in a wide range of disciplines, including economics, consumer behavior, automotive engineering, public policy, and environmental analysis. We developed a semi-structured interview protocol with open-ended questions, asking participants to discuss the strengths and weaknesses of CAFE and several other policy options to reduce fuel consumption by cars and light trucks, particularly market-based incentives that figured prominently in recent legislation and published research. We also asked experts to identify those options that they thought had the greatest potential to reduce fuel consumption, and we discussed how these options could complement or replace the CAFE program. The options selected for discussion in the report represent those alternatives that many of these experts viewed as most promising to reduce fuel consumption by cars and light trucks. To obtain information on policies currently being used to complement CAFE, such as the hybrid tax credit, the Gas Guzzler Tax, and efforts to expand the market for alternative fuels, we relied on our recently published work, relevant legislation, and program information publicly available on government agency Web sites. We conducted our work from August 2006 through June 2007 in accordance with generally accepted government auditing standards.

Appendix II

SELECTED MANUFACTURERS' CAFE PERFORMANCE, SELECTED YEARS FROM 1990 THROUGH 2005

Table 4. BMW CAFE Performance

Year	Domestic car CAFE standard	BMW domestic car CAFE rating	Imported car CAFE standard	BMW imported car CAFE rating	Light truck CAFE standard	BMW light truck CAFE rating
1990	27.5	n/a	27.5	22.2	20.5	n/a
1995	27.5	n/a	27.5	25.3	20.6	n/a
2000	27.5	n/a	27.5	24.8	20.7	17.5
2005	27.5	n/a	27.5	26.9	21.0	21.6

Source: NHTSA.

Table 5. Ford CAFE Performance

Year	Domestic car CAFE standard	Ford domestic car CAFE rating	Imported car CAFE standard	Ford imported car CAFE rating	Light truck CAFE standard	Ford light truck CAFE rating
1990	27.5	26.3	27.5	32.4	20.0	20.2
1995	27.5	27.7	27.5	34.0	20.6	20.8
2000	27.5	28.3	27.5	27.4	20.7	21.0
2005	27.5	28.2	27.5	28.4	21.0	21.5

Source: NHTSA.
Note: Prior to 1991, NHTSA issued separate CAFE standards for two- and four-wheel drive light trucks. The higher figure is used here.

Table 6. General Motors (GM) CAFE Performance

Year	Domestic car CAFE standard	GM domestic car CAFE rating	Imported car CAFE standard	GM imported car CAFE rating	Light truck CAFE standard	GM light truck CAFE rating
1990	27.5	27.1	27.5	32.3	20.0	19.6
1995	27.5	27.4	27.5	36.7	20.6	20.1
2000	27.5	27.9	27.5	25.4	20.7	21.0
2005	27.5	28.8	27.5	29.3	21.0	21.5

Source: NHTSA.

Note: Prior to 1991, NHTSA issued separate CAFE standards for two- and four-wheel drive light trucks. The higher figure is used here.

Table 7. Honda CAFE Performance

Year	Domestic car CAFE standard	Honda domestic car CAFE rating	Imported car CAFE standard	Honda imported car CAFE rating	Light truck CAFE standard	Honda light truck CAFE rating
1990	27.5	n/a	27.5	30.8	20.0	n/a
1995	27.5	n/a	27.5	32.7	20.6	n/a
2000	27.5	31.4	27.5	29.3	20.7	25.4
2005	27.5	36.7	27.5	31.5	21.0	24.8

Source: NHTSA.

Note: Honda did not build domestic cars or any light trucks for the U.S. market in 1990 or 1995.

Table 8. Toyota CAFE Performance

Year	Domestic car CAFE standard	Toyota domestic car CAFE rating	Imported car CAFE standard	Toyota imported car CAFE rating	Light truck CAFE standard	Toyota light truck CAFE rating
1990	27.5	n/a	27.5	30.8	20.5	24.1
1995	27.5	28.5	27.5	30.4	20.6	21.2
2000	27.5	33.3	27.5	28.9	20.7	21.8
2005	27.5	34.3	27.5	35.1	21.0	23.1

Source: NHTSA.

Note: Prior to 1992, NHTSA issued separate CAFE standards for two- and four-wheel drive light trucks. The higher figure is used here. In 1990, Toyota did not build any domestic cars.

REFERENCES

[1] Effectiveness and Impact of Corporate Average Fuel Economy (CAFE) Standards. National Research Council. (Washington, D.C.: 2002).

[2] A cost-effectiveness analysis is used to determine the least-cost option for achieving a specified objective with a given level of benefits. It is one of the commonly used tools to determine whether government investments or programs can be justified on economic principles. These tools also help to identify the best alternative from a range of competing investment alternatives.

[3] Biofuels are a type of alternative fuel made from corn or soybeans or, in the case of cellulosic ethanol from low value agricultural byproducts like cornstalks that are in abundant supply. Alternative fuels include a wider set of fuels that are not made from petroleum, including biofuels, hydrogen, natural gas, and potentially fuels produced by converting coal to liquid, and others. Biofuels offer several environmental advantages, including coming from renewable resources and emitting lower levels of carbon dioxide when they are consumed compared with conventional gasoline and alternative fuels such as those produced from coal.

[4] GAO, Biofuels: DOE Lacks a Strategic Approach to Support Increasing Production with Infrastructure Development and Vehicle Needs, GAO-07-713 (Washington, D.C.: June 8, 2007).

[5] Pub. Law 94-163, codified as positive law at 49 U.S.C. Ch. 329.

[6] A model's CAFE figure generally differs from the window sticker a new vehicle displays showing its fuel economy. The law [49 U.S.C. § 32904(c)] requires that CAFE values be determined through a specific set of test procedures in place at the time EPCA was passed, while window stickers are based on EPA's best estimates of real world fuel economy. Based on the new fuel economy labeling methodology that EPA adopted in 2006, CAFE values are, on average for the industry as a whole, about 25 percent higher than window sticker fuel economy values. CAFE test procedures do not take into account real-world driving conditions such as the use of air conditioning and high-speed driving. EPA officials stated that this results in CAFE figures that are higher than the fuel economy that consumers actually receive from their vehicles.

[7] The Secretary of Transportation issued interim standards for 1981 to 1984.

[8] The Administration submitted similar plans in 2002, 2005, and 2006, but Congress did not act on them.

[9] For example, a manufacturer meets the standard if the average mpg of all the vehicles it manufactures in a model year meet the CAFE standard for that model year. Manufacturers have had to meet an mpg of 27.5 for cars since 1990.

[10] EPCA considers a vehicle to be domestic if at least 75 percent of the cost of the vehicle to the manufacturer is attributable to value added in the United States, Mexico, or Canada. Through rule making, NHTSA required manufacturers to meet the same fleet distinction rule for light trucks, but eliminated it starting in model year 1996. Thus, light truck CAFE standards are calculated as one distinct fleet of a given manufacturer.

[11] 49 C.F.R. § 523.5.

[12] GVWR represents the weight of a vehicle when fully loaded with passengers and cargo.

[13] CAFE penalties are deposited in the U.S. Treasury and are not retained by DOT. The $678 million noted here is not adjusted for inflation.

[14] 26 U.S.C. § 4064.

[15] The number of credits a manufacturer earns is determined by multiplying the tenths of a mpg that the manufacturer exceeded the CAFE standard for a class of vehicles in a model year by the amount of vehicles it manufactured in that class in that model year.

[16] Alternative fuels are fuels or energy sources other than conventional fossil fuels and include ethanol, hydrogen, and batteries.
[17] NHTSA has the authority to continue this credit through rule making.
[18] This conclusion of the NAS report was not unanimous. Two members of the panel that authored the 2002 report dissented from this conclusion. Also, the panel concluded that manufacturers could improve fuel economy while maintaining vehicle weight and that the safety impact of future increases in CAFE standards would depend on many factors. NAS recommended further research on this issue.
[19] The law does not prevent NHTSA from reforming the light truck CAFE program, as it does the car program.
[20] Some experts have noted that if manufacturers shifted their fleet mix toward light trucks with the largest footprints, the average fuel economy of the light truck fleet could decrease from current levels. It is unclear whether complying with the new CAFE standards would cause manufacturers to make larger vehicles, but some experts have suggested that if this became a problem an "antibacksliding" provision could be incorporated in the program to ensure fuel savings. Such a provision would establish a single standard based on the current fleet average below which a manufacturer's fleet could not fall, regardless of compliance with the attribute-based standards.
[21] 71 Fed. Reg. 17566 (2006).
[22] EPCA included a so-called legislative veto provision allowing either the House of Representatives or the U.S. Senate to disapprove any attempt to increase car CAFE standards above the current 27.5 mpg level (or decrease them below 26.0 mpg). However, the Supreme Court has held that this provision is unconstitutional. INS v. Chadha, 462 U.S. 919 (1983). The law does not restrict NHTSA's ability to adjust the light truck CAFE standard or restructure the light truck CAFE program.
[23] This conclusion of the NAS report was not unanimous. Two members of the panel that authored the 2002 report dissented from this conclusion. Also, the panel concluded that manufacturers could improve fuel economy while maintaining vehicle weight and that the safety impact of future increases in CAFE standards would depend on many factors. NAS recommended further research on this issue.
[24] H.R. 6 as amended, 110th Congress.

[25] Some production classified as foreign in 1978 would likely be classified as domestic today, as NHTSA now treats vehicles manufactured in Canada or Mexico as domestically made vehicles.

[26] In this study, NAS assumed no increase in vehicle performance, such as additional horsepower.

[27] In April 2007, the Supreme Court, in a case arising out of EPA's denial of a petition by the state of Massachusetts, among others, ruled that EPA has the authority to regulate greenhouse gas emissions from vehicles and that the agency must either regulate greenhouse gases or explain why it will not or cannot regulate these gases. In denying the petition, EPA officials had stated that one reason they had not issued regulations to reduce greenhouse gas emissions by passenger vehicles was that DOT, not EPA, had the authority to regulate fuel economy, and therefore greenhouse gas emissions, through the CAFE program. However, the Court stated that EPA and DOT could coordinate any rule makings on fuel economy and stated that "there is no reason to think the two agencies cannot both administer their obligations and yet avoid inconsistency."

[28] "Can Proactive Fuel Economy Strategies Help Consumers Mitigate Fuel Price Risks?" University of Michigan Transportation Research Institute (Ann Arbor, Michigan: Sept. 14, 2006); and Espey and Nair, "Automobile Fuel Economy: What Is It Worth?" Contemporary Economic Policy, fall 2005.

[29] NHTSA officials stated that, in addition to the authority the Federal Civil Penalties Inflation Adjustment Act of 1990 under EPCA, NHTSA has the authority to raise CAFE penalties to $10 per 0.1 mpg shortfall.

[30] A 2002 report by DOT, EPA, and DOE estimated that 1 percent of the fuel that flex-fuel vehicles consumed was E85, though this number is likely higher now due to increased availability of E85.

[31] H.R. 6, as amended, 110th Congress.

[32] GAO, 21st Century Challenges: Reexamining the Base of the Federal Government, GAO-05-325SP (Washington, D.C.: February 2005).

[33] GAO, Civil Penalties: Agencies Unable to Fully Adjust Penalties for Inflation under Current Law, GAO-03-409 (Washington, D.C.: Mar. 14, 2003).

[34] GAO-07-713.

[35] The act also created tax credits for purchasing diesel, fuel cell, and dedicated alternative fuel vehicles.

[36] Dierkers, G.; Houdashelt, M.; Silsbe, E.; Stott, S.; Winkelman, S.; & Wubben, M. CCAP Transportation Emissions Guidebook Part Two: Vehicle Technology and Fuels, Center for Clean Air Policy; Washington, D.C. Available online at www.ccap.org/guidebook.

[37] GAO, Government and Performance Accountability: Tax Expenditures Represent a Substantial Federal Commitment and Need to Be Reexamined, GAO-05-690 (Washington, D.C.: Sept. 23, 2005). Also, the Government Performance and Results Act of 1993 requires executive branch agencies to evaluate tax expenditures that affect their missions, and we have noted that outcome-oriented performance goals are important in such evaluations.

[38] One expert estimated that a feebate system that included a rebate of about $2,000 to $2,500 for fuel-efficient vehicles would roughly double demand for these vehicles. Another study estimated that a feebate system paying or charging a minimum of $1,000 could be effective. We did not evaluate the accuracy of these estimates.

[39] On June 11, 2007, the exchange rate was 1 Denmark Kroner = 0.179372 U.S. Dollars.

[40] GAO-05-325SP.

[41] In a system of carbon taxes, each fossil fuel would be taxed, with the tax in proportion to the amount of carbon dioxide released in its combustion. In this and later sections, "carbon" refers to carbon dioxide.

[42] CBO. The Economic Costs of Fuel Economy Standards Versus a Gasoline Tax, December 2003. Washington, D.C. CBO's estimate assumes that manufacturers with high cost of complying with CAFE standards cannot buy "credits" from those that exceeded the standards. Under this assumption gas tax would achieve the targeted reduction in fuel consumption at 19 percent less cost per year compared to increased CAFE standards after all vehicles have been turned over and replaced by vehicles meeting the new CAFE standard. If the credit trading is allowed, CBO estimated that increasing the gas tax would still cost less than increasing CAFE standards but not by as much—about 3 percent annually. CBO's estimates are consistent with what economists told us and the findings of the empirical studies we reviewed. For example, Murphy and Rosenthal, "Allocating the Added Value of Energy Policies" Energy Journal, 2006. Vol. 27, No. 2; pg. 143; Sarah E West,

Roberton C Williams III, "The Cost of Reducing Gasoline Consumption", American Economic Review, 2005. Vol. 95, No. 2; pg. 294-300; David Austin, Terry Dinan, "Clearing the air: The costs and consequences of higher CAFE standards and increased gasoline taxes, "Journal of Environmental Economics and Management, 2005. Vol. 50, No. 3; pg. 562. The studies all found that increasing the tax on gasoline or instituting a tax on carbon is more cost effective than tightening CAFE standards in reducing gasoline consumption.

[43] According to Fischer, Harrington and Parry, "Should Automobile Fuel Efficiency Standards Be Tightened?" Resources for the Future, 2007, the range of the rebound effect is 6 to 10 percent, which is consistent with the estimate Small & Van Dender, "Fuel Efficiency and Motor Vehicle Travel: The Declining Rebound Effect" Energy Journal, No. 28, 2007. However, in its estimation, NHTSA used a range of 10 to 20 percent for rebound effect based on earlier studies.

[44] GAO-07-713.

[45] GAO-07-713.

[46] Ethanol production increased from 3.4 billion gallons in 2004 to 4.9 billion gallons in 2006.

[47] GAO, U.S. Postal Service: Vulnerability to Fluctuating Fuel Prices Requires Improved Tracking and Monitoring of Consumption Information, GAO-07-244 (Washington, D.C.: Feb. 16, 2007).

[48] Pub. Law 108-357.

[49] In addition to GAO's June 2007 report, cited above, see also Congressional Budget Office's discussion of the exemption for alcohol fuels from excise taxes. Congressional Budget Office. Budget Options. (February 2007) Washington, D.C., pp. 324-325.

[50] A current example is the cap-and-trade program for sulfur dioxide under the Clean Air Act. This program includes electric utilities, which are the primary emitters of sulfur dioxide, and established a cap on the utilities' emissions. Sulfur dioxide allowances were primarily given (rather than auctioned) to companies. The program is noteworthy because it represented the first large-scale attempt to set overall emissions levels by using marketable allowances and a choice of compliance methods to control emissions rather than using regulations that specify what actions must be undertaken.

[51] We are currently convening a panel of economists to evaluate the benefits, costs, and trade-offs of climate change policy options. We expect to complete this work in early 2008.

[52] For example, see Resources for the Future (2007). Emissions trading versus CO_2 taxes. Washington, D.C.

INDEX

A

accuracy, 61
administration, vii, 2, 6, 13, 21, 23, 25, 42, 47
advocacy, viii, 6, 53
agricultural, 57
alcohol, 62
alternative (s), viii, 3, 9, 14, 24, 29, 31, 33, 36, 38, 41, 42, 43, 44, 48, 53, 54, 57, 61
American Jobs Creation Act, 44
appendix, ix
appropriations, 2, 13
Asian, 7
assumptions, 30, 53
atmosphere, 17, 44
attention, 51
auditing, ix, 54
authority, 2, 4, 5, 9, 12, 13, 14, 15, 17, 21, 23, 25, 26, 47, 49, 51, 59, 60
automakers, 2, 6
availability, 3, 24, 28, 33, 43, 44, 60
averaging, 5

B

barriers, 35, 43
battery (ies), 34, 59
behavior, 35, 41, 46, 53
benefits, viii, 26, 30, 31, 35, 36, 38, 40, 41, 44, 52, 53, 57, 62
biofuels, 3, 33, 41, 43, 44, 57
burning, 45

C

California, xi, 17, 43
California Air Resources Board (CARB), xi, 17
Canada, 21, 36, 58, 60
capacity, 25, 34
carbon, 3, 17, 31, 33, 39, 40, 41, 43, 44, 45, 46, 48, 52, 57, 61, 62
carbon cap-and-trade program, 33, 45, 48
carbon dioxide, 17, 31, 43, 44, 45, 48, 52, 57, 61
cargo, 7, 27, 58
civil penalties, 4, 7
classes, 2, 9
classification, 27
Clean Air Act, 62
climate change, 17, 20, 48, 62
coal, 43, 46, 57
combustion, 61
community, 29
competitive advantage, 24, 45

complement, viii, 3, 33, 34, 36, 39, 40, 41, 52, 53
compliance, 4, 7, 17, 22, 26, 27, 45, 46, 59, 62
concentration, 29
conditioning, 58
Congress, vii, viii, 2, 4, 5, 6, 12, 13, 15, 17, 21, 23, 25, 26, 29, 34, 46, 47, 49, 58, 59, 60
Congressional Budget Office (CBO), xi, 20, 39, 52, 61, 62
consensus, 31, 48
consulting, 53
consumer choice, 2, 16, 36, 38, 39, 48, 49
consumers, 3, 6, 16, 19, 33, 34, 35, 36, 37, 38, 39, 40, 42, 43, 46, 58
consumption, vii, viii, 1, 2, 3, 4, 5, 6, 9, 12, 14, 15, 17, 20, 25, 33, 38, 39, 40, 41, 42, 43, 45, 47, 48, 51, 52, 53, 61
control, 62
corn, 42, 43, 57
Corporate Average Fuel Economy (CAFE), vii, viii, xi, 1, 2, 3, 4, 5, 6, 7, 8, 9, 10, 11, 12, 13, 14, 15, 16, 17, 19, 20, 21, 22, 23, 24, 25, 26, 27, 28, 29, 30, 33, 34, 36, 39, 40, 41, 42, 43, 46, 47, 48, 49, 51, 52, 53, 55, 56, 57, 58, 59, 60, 61
cost benefit analysis, 14, 51
cost-effective, 3, 4, 14, 20, 33, 35, 39, 40, 42, 45, 48, 57
costs, viii, 2, 15, 24, 26, 27, 28, 30, 35, 41, 43, 44, 45, 46, 47, 53, 62
credit, 3, 4, 13, 20, 27, 28, 34, 35, 36, 44, 45, 46, 48, 49, 51, 59, 61
customers, 22, 42

D

deaths, 11, 16
decisions, 2, 38, 39, 40, 51
deficit, 9
definition, 27
delivery, 44
demand, 19, 20, 24, 34, 40, 43, 44, 61
denial, 60
Denmark, 38, 61
Department of Energy, viii, xi, 52
Department of Energy (DOE) (DOE), viii, xi, 4, 10, 31, 34, 37, 43, 52, 53, 57, 60
Department of Transportation (DOT), vii, xi, 2, 4, 13, 30, 43, 58, 60depreciation, 38, 48
downsizing, 12, 16
draft, 4

E

economic (s), vii, viii, 19, 39, 41, 53, 57
economies of scale, 35
economy (ies), viii, 1, 2, 3, 4, 5, 7, 8, 9, 10, 11, 12, 13, 14, 15, 16, 17, 19, 23, 24, 27, 29, 30, 34, 35, 36, 37, 38, 39, 40, 41, 42, 43, 45, 47, 48, 49, 51, 52, 53, 58, 59, 60
electric utilities, 62
embargo, 5
emission, 31
emitters, 62
employment, 21
energy, vii, viii, 5, 44, 59
Energy Policy Act, 34, 44
Energy Policy and Conservation Act (EPCA), xi, 5, 58, 59, 60
engineering, 30, 53
environmental, vii, viii, 6, 35, 41, 43, 44, 53, 57
environmental advantage, 43, 57
Environmental Protection Agency (EPA), viii, xi, 1, 4, 5, 7, 8, 17, 24, 31, 43, 51, 52, 53, 58, 60
equipment, 24
equity, 2, 23, 48, 49
estimating, 30
ethanol, 24, 42, 43, 44, 57, 59, 62
Europe, 15

European, 7
evidence, 47
exchange rate, 61
excise tax, 62
expenditures, 61
expert (s), viii, 2, 3, 14, 15, 16, 17, 19, 20, 21, 24, 26, 29, 30, 31, 33, 35, 36, 39, 47, 52, 53, 59, 61
expertise, 24, 29, 30, 52, 53

F

fatalities, 10, 13, 23
fear, 30
federal government, 35, 44
fee (s), 36, 38
firms, 53
flex, 24, 29, 43, 44, 51, 60
flexibility, 3, 4, 27, 45, 49
fluctuations, 40
food products, 43
Ford, 40, 55
fossil fuel (s), 46, 59, 61
France, 38
fuel, vii, viii, 1, 2, 3, 4, 5, 7, 8, 9, 10, 11, 12, 13, 14, 15, 16, 17, 19, 20, 22, 23, 24, 26, 27, 28, 29, 30, 33, 34, 35, 36, 37, 38, 39, 40, 41, 43, 44, 45, 47, 48, 49, 51, 52, 53, 57, 58, 59, 60, 61
fuel cell, 61
fuel economy, viii, 2, 3, 5, 8, 10, 11, 12, 13, 15, 17, 19, 24, 27, 30, 34, 35, 36, 37, 38, 39, 40, 42, 47, 48, 51, 52, 58, 60
fuel efficiency, 24
fuel-efficient, 4, 14, 15, 20, 33, 36
fuel-saving, 3, 24, 33, 34, 35, 36, 38, 41, 46, 48
funding, 30, 41, 47
funds, 2, 13

G

gas (es), 17, 31, 60, 61
Gas Guzzler Tax, 3, 8, 36, 37, 38, 54
gasoline, vii, 1, 3, 5, 6, 9, 17, 19, 24, 34, 35, 39, 40, 41, 42, 43, 44, 45, 48, 57, 62
gasoline tax, 39, 41, 42, 48
General Motors (GM), xi, 56
global climate change, 17
Global Positioning System (GPS), xi, 41
goals, 2, 3, 4, 20, 33, 35, 38, 41, 48, 61
government, ix, 4, 31, 35, 36, 39, 42, 44, 45, 53, 57
Government Accountability Office (GAO), 1, 10, 11, 20, 57, 60, 61, 62
grants, 17, 26
greenhouse gas (es), 5, 14, 17, 31, 44, 45, 60
gross vehicle weight rating (GVWR), xi, 7, 12, 58
groups, viii, 6, 52
guidance, viii, 51, 53

H

Honda, 56
House, 2, 6, 28, 44, 59
Hurricane Katrina, 40
hybrid (s), 3, 34, 35, 38, 39, 44, 54
hydrogen, 57, 59

I

Immigration and Naturalization Service (INS), xi, 59
implementation, 26, 52
incentives, viii, 3, 4, 14, 16, 33, 34, 35, 39, 40, 42, 48, 53
income, 39, 42
independence, 44
indexing, 4

industry, viii, 12, 14, 19, 20, 21, 26, 30, 36, 40, 44, 45, 52, 53, 58
inflation, 4, 23, 28, 37, 58
infrastructure, 44
initiation, 52
insurance, viii, 52
internal combustion, 34
Internal Revenue Service (IRS), xi, 8, 36
interpretation, 22
interview (s), 3, 33, 53
investment, 57

J

jobs, 21

L

labeling, 51, 58
labor, 6, 21
land, 43
language, 13
law (s), 5, 13, 22, 39, 44, 47, 51, 57, 58, 59
lead, 4, 13
legislation, 2, 3, 6, 13, 29, 33, 51, 53
light trucks, vii, viii, 1, 2, 3, 4, 5, 6, 7, 9, 11, 12, 13, 15, 16, 17, 19, 20, 21, 23, 25, 27, 29, 33, 36, 37, 40, 41, 43, 45, 47, 48, 51, 52, 53, 55, 56, 58, 59
likelihood, 31
limitation, 36
literature, 39, 52, 53
long-term, 33
low-income, 42

M

maintenance, 41
manufacturer (s), vii, viii, 1, 2, 5, 6, 7, 8, 9, 10, 11, 12, 13, 14, 15, 16, 20, 21, 23, 24, 26, 27, 28, 29, 30, 31, 34, 35, 36, 37, 40, 43, 45, 46, 47, 48, 51, 52, 58, 59, 61
manufacturing, 21, 40
market, viii, 3, 4, 9, 19, 24, 33, 34, 35, 36, 37, 38, 42, 43, 44, 45, 46, 47, 48, 49, 53, 56
market incentives, 48
market share, 9, 19
market-based incentives, 3
marketing, 24
Massachusetts, 60
measures, 35
Mercedes-Benz, 22
Mexico, 21, 58, 60
miles per gallon (MPG), vii, xi
missions, 61
modeling, 30, 53
models, 1, 8, 24, 25, 43, 51, 53

N

nation, vii, 2, 4, 33, 41, 47, 48
National Academy of Sciences (NAS), viii, xi, 2, 10, 13, 14, 15, 21, 23, 29, 30, 47, 49, 52, 53, 59, 60
National Highway Traffic Safety Administration (NHTSA), viii, xi, 1, 2, 4, 5, 6, 7, 8, 9, 10, 11, 12, 13, 14, 15, 16, 17, 21, 23, 24, 25, 26, 27, 28, 29, 30, 37, 47, 49, 51, 52, 53, 55, 56, 58, 59, 60, 62
National Research Council, 57
national security, vii
natural gas, 57
North America, 21, 22, 23
North American Free Trade Agreement, 21

O

Office of Management and Budget, 53

oil, vii, viii, 1, 2, 3, 5, 6, 9, 12, 14, 25, 26, 38, 39, 41, 42, 43, 44, 45, 46, 47, 48, 51, 53
oil consumption, vii, viii, 2, 3, 5, 47, 48
openness, 29
Oregon, 41
organization (s), viii, 30, 52, 53

P

passenger, 4, 5, 7, 8, 9, 10, 15, 17, 27, 37, 39, 44, 48, 49, 60
payroll, 39
peer review, 52
penalty (ies), vii, 1, 4, 6, 7, 8, 9, 14, 15, 21, 28, 36, 48, 52, 58, 60
performance, 7, 19, 22, 34, 35, 60, 61
periodic, 26
petroleum products, 20
pollutants, 41
pollution, 6, 35
President Bush, 28
prices, vii, 6, 15, 19, 34, 35, 40
private investment, 43
procedures, 5, 52, 58
producers, 42, 45, 46
production, 8, 28, 37, 40, 41, 42, 43, 60, 62
program, vii, viii, 1, 2, 3, 4, 5, 6, 7, 9, 10, 11, 12, 14, 15, 16, 17, 19, 20, 23, 24, 25, 26, 28, 29, 30, 31, 33, 34, 35, 36, 41, 44, 45, 46, 47, 48, 49, 51, 52, 53, 59, 60, 62
promote, 44
protocol, 53
proxy, 46
public, 6, 22, 36, 42, 43, 53
pumps, 41

R

range, 12, 15, 16, 17, 31, 33, 34, 35, 38, 39, 40, 45, 48, 52, 53, 57, 62

ratings, 19, 36, 38, 39, 40, 42, 48
real terms, 28
recycling, 42
reduction, 2, 17, 30, 39, 61
reforms, 14
regional, 40
regulation (s), viii, 17, 60, 62
relationship, 30
reliability, 52
renewable resource, 43, 57
researchers, 31
resistance, 42, 48
resources, 4, 24, 49, 53
restructuring, 2, 13
revenue, 35, 36, 42, 44
rivers, 42
rule-making, viii, 6, 29, 51, 53

S

safety, viii, 2, 4, 6, 10, 12, 13, 16, 23, 25, 26, 30, 34, 47, 49, 52, 59
safety experts, viii, 52
sales, 7, 22, 24, 40, 51
sample, 7
savings, 2, 4, 12, 15, 26, 30, 33, 36, 38, 39, 40, 42, 44, 45, 48, 59
science, 30
scientific community, 31
searches, 53
Secretary of Transportation, 58
security, viii
Senate, vii, 2, 3, 6, 13, 25, 28, 29, 44, 48, 59
sensitivity, 31
sites, 54
society, 20, 31, 45
soybeans, 43, 57
speed, 58
sport utility vehicle (SUV), xi, 7, 38
sports, 3, 37
staffing, 52
stakeholders, viii, 6, 14, 19, 30, 53

standards, vii, viii, 1, 2, 4, 5, 6, 7, 9, 10, 11, 12, 13, 14, 15, 16, 17, 19, 20, 21, 23, 24, 25, 27, 28, 29, 30, 31, 36, 39, 40, 41, 42, 43, 46, 47, 48, 49, 51, 52, 53, 55, 56, 57, 58, 59, 61, 62
State of the Union address, 28, 42
storage, 34
strategies, 4, 14, 16, 40, 43
strength, 14
structural reforms, 5
sulfur dioxide, 62
Sun, 23
supply, 34, 42, 43, 57
Supreme Court, 59, 60
surplus, 9

T

targets, 6, 12, 13
tax credit (s), viii, 3, 34, 35, 44, 48, 54, 61
tax incentive, 38, 40
tax increase, 37
taxes, 3, 4, 15, 36, 41, 48, 61, 62, 63
technological, 30, 42, 47
technology, 14, 15, 34, 35, 36, 40, 42, 45, 47, 49
test procedure, 58
Texas, 43
time, 2, 4, 12, 14, 20, 23, 26, 33, 35, 38, 40, 45, 46, 47, 48, 58
Toyota, 56
tracking, 51
trade, viii, 3, 4, 27, 29, 33, 44, 45, 46, 48, 52, 62
trade union, viii, 52
trade-off, 62
trading, 3, 4, 9, 13, 20, 27, 45, 46, 48, 49, 61, 63
transition, 12
transportation, vii, 5, 17, 42, 43, 46
Treasury, 9, 44

trucks, viii, 1, 2, 5, 6, 7, 9, 11, 12, 13, 15, 16, 17, 19, 25, 27, 33, 37, 47, 48, 51, 54
trust fund, 41

U

U.S. Treasury, 58
unions, 6
United Auto Workers (UAW), viii, xi, 21, 52
United Kingdom, 38
United States, vii, xi, 3, 5, 9, 15, 20, 33, 36, 38, 43, 44, 47, 58
United States Postal Service (USPS), xi, 44
universities, viii, 53
updating, 14, 26, 47, 48, 52

V

values, 5, 31, 51, 53, 58
variables, 51, 53
vehicle fleet, 10, 19
vehicle miles traveled (VMT), xi, 19, 39, 41
vehicles, vii, 1, 3, 4, 5, 7, 9, 12, 13, 14, 15, 16, 19, 20, 22, 23, 24, 27, 29, 33, 34, 35, 36, 37, 38, 39, 40, 41, 42, 43, 44, 46, 48, 51, 58, 59, 60, 61
visible, 48
Volkswagen, 8, 23

W

Washington, 57, 60, 61, 62, 63
weakness, 20, 35
workers, viii, 21, 52
write-offs, 3, 38, 48